Semantic Mining
of Social Networks

Synthesis Lectures on the Semantic Web: Theory and Technology

Editors
Ying Ding, *Indiana University*
Paul Groth, *Elsevier Labs*

Founding Editor Emeritus
James Hendler, *Rensselaer Polytechnic Institute*

Synthesis Lectures on the Semantic Web: Theory and Application is edited by Ying Ding of Indiana University and Paul Groth of Elsevier Labs. Whether you call it the Semantic Web, Linked Data, or Web 3.0, a new generation of Web technologies is offering major advances in the evolution of the World Wide Web. As the first generation of this technology transitions out of the laboratory, new research is exploring how the growing Web of Data will change our world. While topics such as ontology-building and logics remain vital, new areas such as the use of semantics in Web search, the linking and use of open data on the Web, and future applications that will be supported by these technologies are becoming important research areas in their own right. Whether they be scientists, engineers or practitioners, Web users increasingly need to understand not just the new technologies of the Semantic Web, but to understand the principles by which those technologies work, and the best practices for assembling systems that integrate the different languages, resources, and functionalities that will be important in keeping the Web the rapidly expanding, and constantly changing, information space that has changed our lives.
Topics to be included:

- Semantic Web Principles from linked-data to ontology design

- Key Semantic Web technologies and algorithms

- Semantic Search and language technologies

- The Emerging "Web of Data" and its use in industry, government and university applications

- Trust, Social networking and collaboration technologies for the Semantic Web

- The economics of Semantic Web application adoption and use

- Publishing and Science on the Semantic Web

- Semantic Web in health care and life sciences

Semantic Mining of Social Networks

Jie Tang and Juanzi Li

ISBN: 978-3-031-79461-2   paperback
ISBN: 978-3-031-79462-9   ebook

DOI 10.1007/978-3-031-79462-9

A Publication in the Springer series
SYNTHESIS LECTURES ON THE SEMANTIC WEB: THEORY AND TECHNOLOGY

Lecture #11
Series Editors: Ying Ding, *Indiana University*
          Paul Groth, *Elsevier Labs*
Founding Editor Emeritus: James Hendler, *Rensselaer Polytechnic Institute*
Series ISSN
Print 2160-4711   Electronic 2160-472X

Semantic Mining
of Social Networks

Jie Tang and Juanzi Li
Tsinghua University

SYNTHESIS LECTURES ON THE SEMANTIC WEB: THEORY AND TECHNOLOGY #11

ABSTRACT

Online social networks have already become a bridge connecting our physical daily life with the (web-based) information space. This connection produces a huge volume of data, not only about the information itself, but also about user behavior. The ubiquity of the social Web and the wealth of social data offer us unprecedented opportunities for studying the interaction patterns among users so as to understand the dynamic mechanisms underlying different networks, something that was previously difficult to explore due to the lack of available data.

In this book, we present the architecture of the research for social network mining, from a microscopic point of view. We focus on investigating several key issues in social networks. Specifically, we begin with analytics of social interactions between users. The first kinds of questions we try to answer are: What are the fundamental factors that form the different categories of social ties? How have reciprocal relationships been developed from parasocial relationships? How do connected users further form groups?

Another theme addressed in this book is the study of social influence. Social influence occurs when one's opinions, emotions, or behaviors are affected by others, intentionally or unintentionally. Considerable research has been conducted to verify the existence of social influence in various networks. However, few literature studies address how to quantify the strength of influence between users from different aspects. In Chapter 4 and in [138], we have studied how to model and predict user behaviors. One fundamental problem is distinguishing the effects of different social factors such as social influence, homophily, and individual's characteristics. We introduce a probabilistic model to address this problem.

Finally, we use an academic social network, ArnetMiner, as an example to demonstrate how we apply the introduced technologies for mining real social networks. In this system, we try to mine knowledge from both the informative (publication) network and the social (collaboration) network, and to understand the interaction mechanisms between the two networks. The system has been in operation since 2006 and has already attracted millions of users from more than 220 countries/regions.

KEYWORDS

social tie, strong/weak ties, parasocial interactions, reciprocity, social influence, collective classification, graphical model, social network analysis, social relationship, relationship mining, link prediction, influence maximization, network centrality, user modeling, social action, social theories, social balance, social status, triadic closure, factor graph, influence propagation, conservative influence propagation, non-conservative influence propagation, user behaviour prediction, profile extraction, expert finding, name disambiguation, ArnetMiner

Contents

Acknowledgments

This work is supported by the National High-Tech R&D Program (No. 2014AA015103), National Basic Research Program of China (No. 2014CB340506, No. 2012CB316006), Natural Science Foundation of China (No. 61222212, No.61035004), the Tsinghua University Initiative Scientific Research Program (20121088096), a research fund supported by Huawei Inc., and Beijing key lab of networked multimedia.

Jie Tang and Juanzi Li
April 2015

CHAPTER 1

Introduction

A social network is a social structure made up of a set of actors (such as individuals or organizations) and a complex set of the dyadic ties between these actors. Social network mining aims to provide a comprehensive understanding of global and local patterns, mechanism of the network formation, and dynamics of user behaviors. Social network analysis and mining is an inherently interdisciplinary academic field which emerged from sociology, psychology, statistics, and graph theory. However, due to the lack of efficient computational models and the nonavailability of large-scale social networking data, traditional research on social networks has mainly focused on qualitative study in small-scale networks. For example, Milgram spent many years validateng the existence of small-world phenomenon, also referred to as six-degrees of separation by sending mail to thousands of people [107]. In the 1910s, Georg Simmel proposed the concept of structural theories in sociology, which focuses on the dynamic formation of triads [132] and in the 1930s, Jacob Moreno was the first to develop sociograms to analyze people's inter-relationships. Later, for quantitatively analyzing social networks, researchers gave mathematical formulations for social network analysis and developed computational methods [157]. However, most social network research was still conducted by interviewing the participants with small-scale social networks.

More recently, with the emergence and rapid proliferation of online social applications and media—such as instant messaging (e.g., IRC, AIM, MSN, Jabber, Skype), sharing sites (e.g., Flickr, Picassa, YouTube, Plaxo), blogs (e.g., Blogger, WordPress, LiveJournal), wikis (e.g., Wikipedia, PBWiki), microblogs (e.g., Twitter, Jaiku, Weibo), social networks (e.g., Facebook, MySpace, Ning), collaboration networks (e.g., DBLP, ArnetMiner), to mention a few—these online services bring many opportunities for studying social networks, while also posing a number of new challenges. First, the online social network is much larger than the physical social networks in traditional research. Facebook had more than 1.3 billion users in 2014 and Tencent, the largest social networking service in China, attracted 800 million users in 2014. Twitter, the largest microblogging service, has hit half a billion tweets a day. All these big numbers bring new challenges and also the requirements for developing new methods to store, search, analyze, and mine the "big social data." Second, from a microscopic viewpoint, what are the subtle changes of behaviors when users go online from the offline physical networks? Are the traditional social theories still applicable?

Existing research related to social network mining can be categorized into micro-level, meso-level, and macro-level, although the levels of mining are not necessarily mutually exclu-

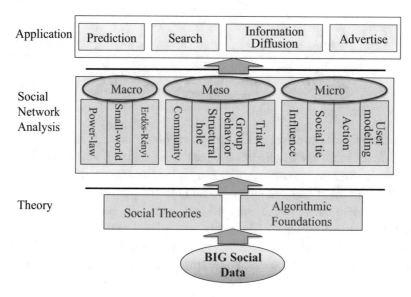

Figure 1.1: Research roadmap for social network mining.

sive. At the micro-level, social network mining is mainly concerned with modeling individuals' behaviors and interactions between users; at the macro-level, it focuses on studying the global patterns of the social networks, for example, network topology [47] and network generative models [10, 46, 158]; and at the micro-level, social network mining is mainly concerned with modeling individuals' behaviors and interactions between users, for example social action theory [159], social ties [56], social influence [79, 143], and user modeling [48]. The meso-level research falls between the micro- and macro-levels, for example community detection [53, 113, 114], structural hole [21, 100], and group behavior analysis [4, 34]. Figure 1.1 gives the research roadmap for social network mining. On top of social network mining, we can consider many applications such as social prediction [138], social search [41, 148], information diffusion [59], and social advertisement [9]. The underlying theories for social network mining include theories from social science and algorithmic foundations from computer science.

In this book, we study social network mining from a microscopic viewpoint. We focus on modeling users' behaviors and their interactions between each other. In particular, we present key technologies for social tie analysis, social influence analysis, and user behavior modeling.

1.1   BACKGROUND

We introduce several related social theories and then briefly review existing literatures on social tie analysis, social influence analysis, and user behavior modeling.

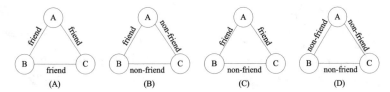

Figure 1.2: Illustration of structural balance theory. (A) and (B) are balanced, while (C) and (D) are not balanced.

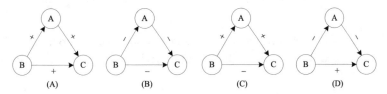

Figure 1.3: Illustration of status theory. (A) and (B) satisfy the status theory, while (C) and (D) do not satisfy the status theory. Here positive "+" denotes the target node has a higher status than the source node; and negative "-" denotes the target node has a lower status than the source node. In total there are 16 different cases.

1.1.1   SOCIAL THEORIES

A basic principle for mining social networks in this book is to incorporate social theories into data mining (or machine learning) model. For social theories, we mainly consider social balance [45], social status [34], structural hole [21], two-step information-flow [89], and strong/weak tie hypothesis [56, 83].

Social balance theory suggests that people in a social network tend to form into a balanced network structure. Figure 1.2 shows such an example to illustrate the structural balance theory over triads, which is the simplest group structure to which balance theory applies. For a triad, the balance theory implies that either all three of these users are friends—"the friend of my friend is my friend"— or only one pair of them are friends—"the enemy of my enemy is my friend."

Another social psychological theory is the *theory of status* [34, 60, 93]. This theory is based on the directed relationship network. Suppose each directed relationship is labeled by a positive sign "+" or a negative sign "-" (where sign "+"/"-" denotes the target node has a higher/lower status than the source node). Then status theory posits that if, in a triangle on three nodes (also-called triad), we take each negative edge, reverse its direction, and flip its sign to positive, then the resulting triangle (with all positive edge signs) should be acyclic. Figure 1.3 illustrates four examples. The first two triangles satisfy the status ordering and the latter two do not satisfy it.

Roughly speaking, a user is said to span a *structural hole* in a social network if she is linked to people in parts of the network that are otherwise not well connected to one another [21]. Such a user is also referred to as *structural hole spanner* [100]. Arguments based on structural holes suggest that there is an informational advantage to have friends in a network who do not know

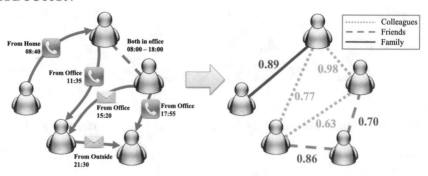

Figure 1.4: An example of inferring social ties in a mobile communication network. The left figure is the input of the task, and the right figure is the output of the task of inferring social ties.

each other. A sales manager with a diverse range of connections can be considered as a structural hole spanner, with a number of potentially *weak ties* [56] to individuals in different communities. More generally, we can think about websites, such as eBay, as spanning structural holes, in that they facilitate economic interactions between people who would otherwise not be able to find each other.

The *two-step information-flow theory* is first introduced in [89] and further elaborated in literature [75, 76]. The theory suggests that ideas (innovations) usually flow first to *opinion leaders*, and then from them to a wider population. In the enterprise email network, for example, managers may act as opinion leaders to help spread information to subordinates.

Interpersonal ties generally come in three varieties: strong, weak, or absent. *Strong tie hypothesis* implies that one's close friends tend to move in the same circles that she/he does, while *weak tie hypothesis* argues that weak ties are responsible for the majority of the embeddedness and structure of social networks in society as well as the transmission of information through these networks [56].

1.1.2   SOCIAL TIE ANALYSIS

Mining social ties is an important problem in social network analysis. Based on the strong/weak tie hypothesis, there is a bunch of research conducted in recent years. The goal of *social tie analysis* is to automatically recognize the semantics associated with each social relationship. Awareness of the semantics of social relationships can benefit many applications. For example, if we could have extracted friendships between users from the mobile communication network, then we can leverage the friendships for a "word-of-mouth" promotion of a new product. Figure 1.4 gives an example of relationship mining in a mobile calling network. The left figure is the input of the problem: in a mobile calling network consisting of users, calls and messages between users, and users' location logs, etc. The objective is to infer the type of the relationships in the network. In the right figure, the users who are family members are connected with red-colored lines, friends are connected with blue-colored dash lines, and colleagues are connected with green-colored dotted lines. The probability associated with each relationship represents our confidence on the detected relationship types.

There are several works on mining the relationship semantics. Diehl et al. [38] tried to identify the manager-subordinate relationships by learning a ranking function. They defined a ranking objective function and cast the relationship identification as a relationship ranking problem. Menon et al. [106] proposed a log-linear matrix model for dyadic prediction. They used matrix factorization to derive latent features and incorporate the latent features for predicting the label of user relationships. Wang et al. [155] proposed a probabilistic model for mining the advisor-advisee relationships from the publication network. The proposed model is referred to as time-constrained probabilistic factor graph model (TFGM), which supports both supervised and unsupervised learning. Eagle et al. [44] presented several patterns discovered in mobile phone data, and try to use these pattern to infer the friendship network. Tang et al. [149] developed a classification framework of social media based on differentiating different types of social connections. However, these algorithms mainly focus on a specific domain, while our model is general and can be applied to different domains. Moreover, these methods also do not explicitly consider the correlation information between different relationships.

Another research branch is to predict and recommend unknown links in social networks. Liben-Nowell et al. [96] studied the problem of inferring new interactions among users given a snapshot of a social network. They developed several unsupervised approaches to deal with this problem based on measures for analyzing the "proximity" of nodes in a network. The principle is mainly based on similarity of either content or structure between users. Backstrom et al. [6] proposed a supervised random walk algorithm to estimate the strength of social links. Leskovec et al. [92] employed a logistic regression model to predict positive and negative links in online social networks, where the positive links indicate the relationships such as friendship, while negative indicates opposition. However, these works consider only the black-white social networks, and do not consider the types of the relationships.

Recently, Hopcroft et al. [70] explored the problem of reciprocal relationship prediction and Lou et al. [101] extended to study how social relationships develop into triadic closure. They proposed a learning framework to formulate the problem of reciprocal relationship prediction into a graphical model and evaluate the proposed method on a Twitter data set. The framework is demonstrated to be very effective, i.e., it is possible to accurately infer 90% of reciprocal relationships in a dynamic network. Tang et al. [142] further proposed a general framework for classifying the type of social relationships by learning across heterogeneous networks. The idea is to use social theories (e.g., social balance theory, social status theory, structural hole theory, two-step flow theory, and strong/weak tie) as bridge to connect different social networks. Social theory-based features are defined and incorporated into a triad-based factor graph model to infer the type of social relationships in different networks.

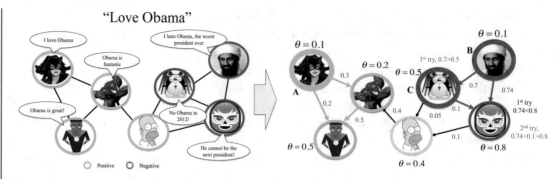

Figure 1.5: An example of social influence for political mobilization. The left figure is the input of the task, and the right figure is the output: influence probability between users, individual conformity of each user, and key influencers (A, B, C).

Another related research topic is relational learning [23, 51]. However, the problem presented in this book is very different. Relational learning focuses on the classification problems when objects or entities are presented in relations, while social tie analysis focuses on exploring the relationship types in social network. A number of supervised methods for link prediction in relational data have been developed [118, 151].

1.1.3   SOCIAL INFLUENCE ANALYSIS

Social influence occurs when one's opinions, emotions, or behaviors are affected by others, intentionally or unintentionally [78]. Recently, social influence analysis has attracted a lot of interest from both research and industry communities. In general, existing research on social influence analysis can be classified into three categories: influence test, influence measure, and influence diffusion models. Figure 1.5 shows an example of social influence for political mobilization. The left figure is the input of the task: opinion of each user for "Obama" in the social network, and the right figure is the output: influence probability between users on this topic "Obama," individual conformity of each user, and key influencers (A, B, C).

Influence Test. Several efforts have been made for identifying the existence of the social influence in the online social networks. For example, Anagnostopoulos et al. [2] gave a theoretical justification to identify influence as a source of social correlation when the time series of user actions is available. They propose a shuffle test to prove the existence of the social influence. Singla and Richardson [133] studied the correlation between personal behaviors and their interests. They found that in the online system people who chat with each other (using instant messaging) are more likely to share interests (their Web searches are the same or topically similar), and the more time they spend talking, the stronger this relationship is. Bond et al. [18] reported results from a randomized controlled trial of political mobilization messages delivered to 61 million Facebook users. They found that when one is aware that their friends have made the political votes, their likelihood to vote will significantly increase. Crandall et al. [31] further investigated the correlation between social similarity and influence. More recently, some efforts have been made for analyzing the dynamics in the social network. For example, Scripps et al. [128] investigated how different pre-processing decisions and different network forces such as selection and influence affect the modeling of dynamic networks. Other similar work can be referred to Dourisboure et al. [43].

Influence Measure. The goal of influence measure is to quantify the strength of influence between users. Tang et al. [143] introduced the problem of topic-based social influence analysis. They proposed a Topical Affinity Propagation (TAP) approach to describe the problem via using a graphical probabilistic model. However, these works neither consider heterogeneous information nor learn topics and influence strength jointly. Goyal et al. [55] and Saito et al. [124] measured the pairwise influence between two individuals based on the idea of independent cascade model [79]. Liu et al. [99] also studied the problem of measuring the influence on different topics. They proposed a generative graphical model which leverages both heterogeneous link information and textual content associated with each user in the network to mine topic-level influence strength. Based on the learned direct influence, we further study the influence propagation and aggregation mechanisms: conservative and non-conservative propagations to derive the indirect influence. Xin et al. [131] studied the indirect influence using the theory of quantum cognition. Myers et al. [112] proposed a probabilistic model to quantify the external influence out-of-network sources. Belak et al. [12] investigated and measured the influence on the cross-community level so as to provide a coarse-grained picture of a potentially very large network. They presented a framework for cross-community influence analysis and evaluated the proposed method on a ten-year data set from the largest Irish online discussion system Boards.ie, Boards.ie is the name of the system. Zhang et al. [174] proposed the notion of social influence locality and applied it for modeling users' retweeting behaviors in the social networks. They developed two instantiation functions based on pairwise influence and structural diversity.

Influence Diffusion Models. Social influence has been applied in the application of influence maximization in viral marketing. Domingos and Richardson [42, 121] were the first to study influence maximization as an algorithmic problem. Kempe et al. [79] took the first step to formulating influence maximization as a discrete optimization problem. Leskovec et al. [94] and Chen et al. [26, 27] made efforts to improve the efficiency of influence maximization. Gruhl et al. [59] proposed a time-decayed diffusion model for blogging writing, and use an EM-like algorithm to estimate the influence probabilities. Yang et al. [167] studied the interplay between users' social roles and their influence on information diffusion. They proposed a Role-Aware INformation diffusion model (RAIN) that integrates social role recognition and diffusion modeling into a unified framework.

1.1.4   USER MODELING AND ACTIONS

User modeling describes the process of building up a user model to characterize users' skills, declarative knowledge, and specific needs to a system [48].

A number of models have been proposed to model users' behaviors in dynamic social networks. Sarkar et al. [125] developed a generalized model associating each entity in Euclidean latent space and used kernel functions for similarity in latent space to model friendship drifting over time. Tan et al. [138] studied how users' behaviors (actions) in a social network are influenced by various factors such as personal interests, social influence, and global trends. They proposed a Noise Tolerant Time-varying Factor Graph Model (NTT-FGM) for modeling and predicting social actions, which simultaneously models social network structure, user attributes, and user action history for better prediction of the users' future actions. Tan et al. [137] investigated how users' sentiment can be inferred in the social network by incorporating the social network information. Scripps et al. [128] presented a model to investigate how different pre-processing decisions and different network forces such as selection and influence affect the modeling of dynamic networks. They also demonstrated the effects of attribute drifting and the importance of individual attributes in forming links over time.

Group analysis is based on the view that deep lasting change can occur within a carefully formed group whose combined membership reflects the wider norms of society. There is an interest, in group analysis, on the relationship between the individual group member and the rest of the group resulting in a strengthening of both, and a better integration of the individual with his or her community, family and social network. Shi et al. [130] studied the pattern of user participation behavior, and the feature factors that influence such behavior on different forum data sets. Backstrom et al. [5] proposed a partitioning on the data that selects for active communities of engaged individuals.

1.1.5   GRAPHICAL MODELS

From the perspective of machine learning and data mining, for analyzing networking data, graphical probabilistic models are often employed to describe the dependencies between observation data. Bayesian networks [72], Markov random field [134], factor graph [50, 84], Restricted Boltzmann Machine (RBM) [162], deep neural networks [68], and many others are widely used graphical models.

1.2   BOOK OUTLINE

Despite much work having been conducted for mining social networks, we are still facing a number of challenges. In this book, from the microscopic viewpoint, we introduce key technologies for social tie analysis, social influence analysis, and user behavior modeling. Figure 1.6 gives an overview of the topics covered in this book and their relationships. Specifically, the book is organized as follows.

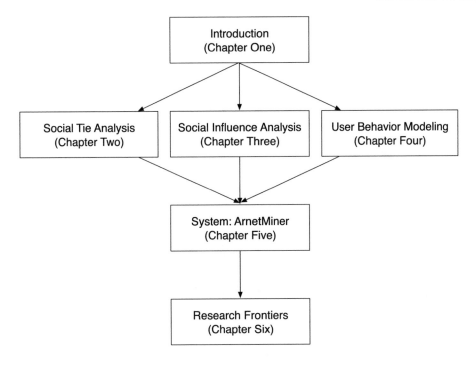

Figure 1.6: Outline of this book.

- In Chapter 2, we begin with the analytics of social interactions between users. The first question we try to answer is: "What are the fundamental factors that form the different categories of social ties?." We incorporate social network theories into a probabilistic graphical model for learning to automatically infer the types of social relationships.

- In Chapter 3, we investigate the problem of social influence. We introduce theoretical methodologies for verifying the existence of social influence, and for quantifying the influence strength between users on different topics.

- In Chapter 4, we study how to model and predict user behaviors. One fundamental problem is to distinguish the effects of different social factors such as social influence, homophily, and individual's characteristics. We introduce a probabilistic model to address this problem.

- In Chapter 5, we use ArnetMiner, an academic social network developed by the authors, as an example to demonstrate how we apply the introduced technologies for mining real social networks at the micro-level.

- Finally, in Chapter 6, we present potential future directions in this field.

CHAPTER 2

Social Tie Analysis

Social ties, also referred to as interpersonal ties, are defined as information-carrying connections between people. In general, from a sociological perspective, social ties can be categorized into three varieties: strong, weak, or absent.[1] From the computational perspective, related research on social tie analysis includes: predicting missing links, inferring social ties, and predicting reciprocity.

2.1 OVERVIEW

In social networks, the basis for social tie analysis is predicting missing links. Liben-Nowell and Kleinberg [96] systematically investigated the problem of inferring new links among users given a snapshot of a social network. They introduced several unsupervised approaches to deal with this problem based on "proximity" of nodes in a network—or the principle of homophily [88] ("birds of a feather flock together" [105]). Besides predicting new links, another important topic in social tie analysis is to automatically recognize the semantics associated with each social relationship, referred to as *inferring social ties*. In online social networks, most relationships are not meaningfully labeled (e.g., "colleague" and "intimate friends"), simply because users do not want to label them in order to protect privacy. Statistics show that only 16% of mobile phone users in Europe have created custom contact groups [58, 123] and less than 23% connections on LinkedIn have been labeled. Awareness of the types of social relationships can benefit many applications. For example, if we could have extracted friendships between users from the mobile communication network, we can leverage the friendships for a "word-of-mouth" promotion of a new product [79]. Figure 2.1 shows the major research topics on social tie analysis. In addition to predicting missing links and inferring social ties, another important topic to understand is how a social reciprocal (two-way) relationship has been developed from a parasocial (one-way) relationship, and why. The problem is called *reciprocity*. In this chapter, we will focus on introducing the first two problems: predicting missing links and inferring social ties. For predicting reciprocity, the interested reader should refer to Hopcroft et al. [70] and Lou et al. [101].

To begin with, let us define the general input of the problems we will address in this chapter. Basically, we are given a social network $G = (V, E)$, where V is a set of $|V| = N$ users and $E \subset V \times V$ is a set of $|E| = M$ relationships between users. Let $e_{ij} \in E$ denote a directed relationship from user v_i to user v_j. In a undirected network, we have $e_{ij} = e_{ji}$. In this case, we use either e_{ij} or e_{ji} to represent the relationship between v_i and user v_j.

---

[1]http://en.wikipedia.org/wiki/Interpersonal_ties

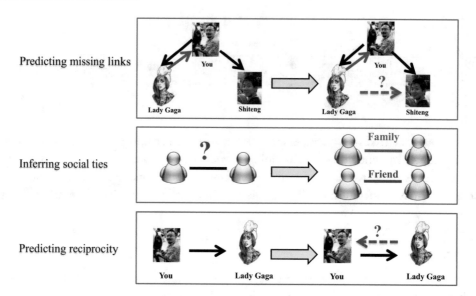

Figure 2.1: Research on social tie analysis.

2.2   PREDICTING MISSING LINKS

In this section, we conducted a survey of existing approaches for link prediction in social networks. As shown in Table 2.1, we classify those approaches into several groups.

Similarity Metric-Based Methods. Similarity metric-based methods compute a similarity score between a pair of nodes (users) and predict the missing links based on the obtained similarity scores—a higher similarity score results in a higher probability to create a link.

Liben-Nowell and Kleinberg [96] proposed a similarity metric-based method to solve the link prediction problem. Given a social network, they first computed the similarity between a pair of vertices by various graph-based similarity metrics and then used the similarity scores to predict the missing link between two vertices. Kashima et al. presented a parameterized probabilistic model of network evolution based on the topological features of network structures for link prediction [74]. Schifanella et al. [127] showed semantic similarity measures among users based solely on their social media annotation metadata are predictive of social links. Richard et al. proposed finding the missing links using the dynamics of the graph through a set of topological features, such as the degrees of the vertices; for additional information see [120]. Clauset et al. [29] showed that knowledge of hierarchical structure can be used to predict missing connections in partly known networks with high accuracy. In [29] they presented the computationally less expensive features based on social vector clocks to solve the link prediction problem.

Table 2.1: Survey of major methodologies for missing link prediction

Category	Method	Description
	LPP [96]	Similarity Survey
	PPM [74]	Topological Features
	FIF [127]	Features for Folks in Folksonomies
	GFT [120]	Graph Feature Tracking
	HSLP [29]	Hierarchical Structure-based Similarity
	SVC [90]	Social Vector Clocks
Similarity Metrics	VCP [97]	Vertex Collocation Profile
	DLP [35]	Local and Global Properties
	LP-MPSN [156]	Features for Mobile Phone Social Network
	LP-LSN [126]	Features for Location-based Social Network
	CTLP [115]	Dimension Reduction to Latent Features
	NLFM [177]	Latent Feature Model
	LP-PS [117]	Popularity vs. Similarity
	LP-SSN [98]	Similarity-based Method for Sparse Network
Matrix Factorization	SGT [86]	Spectral Graph Transformations
	NMD [85]	Nondiagonal Matrix Decompositions
	LPGM [154]	Local Probabilistic Graphical Model
	SBM [61]	Stochastic Block Models
	TPFG [155]	Time-Constrained Probabilistic Factor Graph Model
	CSLP [91]	Bootstrap Probabilistic Graph Model
Graphical Models	LINKREC [172]	Random Walk
	SRW [6]	Supervised Random Walk
	DBN [103]	Dynamic Bayesian Network
	MTLM [178]	Mixed-Topic Link Models
	LFPM [176]	Latent Friendship Propagation Model
	PFGM [164]	Factor Graph Model
	DLFPM [67]	Dynamic Latent Feature Propagation Model

Lichtenwalter et al.[97] introduced the concept of a vertex collocation profile (VCP) for the purpose of topological link analysis and prediction. VCPs provide nearly complete information about the surrounding local structure of embedded vertex pairs. De et al. combined the global properties (graph conductance, hitting or commute times, Katz score), local properties (Adamic-Adar or node feature vectors), and the link densities at the intermediate level of communities into a discriminative link predictor [35]. Wang et al. [156] and Scellato et al. [126] presented different features for link prediction in mobile phone social networks and location-based social networks, respectively. Oyama et al. proposed a dimension reduction approach to cross-temporal link prediction by jointly learning a set of feature projection matrices from the training data [115]. Zhu [177] presented a max-margin nonparametric latent feature model to discover discriminative latent features for link prediction and to automatically infer the unknown latent social dimension.

Papadopoulos et al. showed that popularity is just one dimension of attractiveness; another dimension is similarity. They developed a framework in which new connections optimize certain trade-offs between popularity and similarity, which predicts the probability of new links with high precision in the technological, biological, and social networks [117]. Lichtenwalter et al. [98] explored many factors significant in influencing and guiding classification and presented an effective flow-based predicting algorithm which offers formal bounds on imbalance in sparse network link prediction.

Matrix Factorization-Based Methods. Matrix factorization-based methods model the social network as a matrix and solve the problem of link prediction using the matrix decompositions from linear algebra or graph theories.

Kunegis and Lommatzsch [86] presented a unified framework for learning link prediction and edge weight prediction functions in large networks, based on the transformation of a graph's algebraic spectrum. Their approach generalizes several graph kernels and dimensionality reduction methods and provides a method to estimate the parameters efficiently. Kunegis and Fliege [85] also presented a method for trust prediction based on nondiagonal decompositions of the asymmetric adjacency matrix of a directed network. They used a nondiagonal decomposition into directed components (DEDICOM) to learn the coefficients of a matrix polynomial of the network's adjacency matrix. Their method can be used to compute better low-rank approximations to a polynomial of a network's adjacency matrix than using the singular value decomposition.

Probabilistic Graph Model-Based Methods. The probabilistic graph model based methods model the joint-probability among the nodes by Bayesian graphical models, which are usually complex to mine the latent relations inside the network.

Wang et al. [154] introduced a local probabilistic graphical model method that can scale to large graphs to estimate the joint co-occurrence probability of two nodes. Guimera and Sales Pardo [61] presented a general mathematical and computational framework to deal with the problem of data reliability in complex networks based on stochastic block models, by which both missing and spurious interactions could be reliably identified in noisy network observations.

Wang et al. proposed a time-constrained probabilistic factor graph model (TPFG), which takes a research publication network as input and models the advisor-advisee relationship mining problem using a jointly likelihood objective function [155]. Leroy et al. [91] proposed a two-phase method based on the bootstrap probabilistic graph. The first phase generates an implicit social network under the form of a probabilistic graph. The second phase applies probabilistic graph-based measures to produce the final prediction. Yin et al. [172] presented estimating the link relevance using a random walk algorithm on an augmented social graph with both attribute and structure information. Backstrom et al. developed an algorithm based on supervised random walks that naturally combines the information from the network structure with node- and edge-level attributes [6].

Mathur et al. [103] utilized a dynamic Bayesian network to detect interaction links in a collaborating group using manually annotated data. Zhu et al. combined classic ideas in topic modeling with a variant of the mixed-membership block model, and proposed the mixed-topic link models for unsupervised topic classification and link prediction [178]. Zhang et al. modeled link formation as results of individuals' friend-making behaviors combined with personal interests [176]. They proposed the Latent Friendship Propagation Network (LFPN) to depict the evolution progress of one's egocentric network, and modeled individuals' social behaviors using the Latent Friendship Propagation Model (LFPM). In [164], Wu et al. proposed an interactive learning framework to formulate the problem of recommending patent partners into a factor graph model. The framework involves three phases: candidate generation, candidate refinement, and interactive learning method to efficiently update the existing recommendation model based on inventors' feedback. Heaukulani et al. proposed a latent feature propagation model for link prediction by capturing how observed social relationships from the past affect future unobserved structure in the network [67].

In the following section, we first introduce some different kinds of approaches for missing link prediction.

2.2.1 SIMILARITY METRICS

We first introduce different similarity metrics for link prediction, including the node neighborhood based metrics, path-based metrics, node and edge attributes-based metrics, and latent metrics. Then we introduce the primary challenge for link prediction as classification. Most of the similarity features have been surveyed in Al Hasan and Zaki [64].

Node Neighborhood-based Metrics

Common Neighbors. For two users (nodes), u and v, the number of their common neighbors is denoted as $|\Gamma(u) \cap \Gamma(v)|$, where $\Gamma(\cdot)$ represents the set of neighbors of a node. The idea of using the size of common neighbors is just an attestation to the network transitivity property. In other words, if user u and user v are connected by many common nodes, then it is very likely that user u and v are (or will be) connected.

Jaccard Coefficient. The common neighbors metric will bias toward high-degree users. Jaccard Coefficient is a metric to address this problem by normalizing the number of common neighbors:

$$Jaccard - coefficient(u, v) = \frac{|\Gamma(u) \cap \Gamma(v)|}{|\Gamma(u) \cup \Gamma(v)|}. \tag{2.1}$$

Conceptually, it defines the probability that a common neighbor of a pair of nodes u and v would be selected if the selection is made randomly from the union of the neighbor-sets of u and v. Thus, for a high number of common neighbors, the score would be higher.

Adamic-Adar. Adamic and Adar [1] proposed a new score as a metric of similarity between two nodes. It is defined as below:

$$Adamic - Adar(u, v) = \sum_z \frac{1}{\log \text{freq}(z)} \tag{2.2}$$

where z denotes any feature shared by u and v, and freq(z) denotes the frequency of feature occurring between u and v. When considering common neighbors as features, then the metric can be re-written as:

$$Adamic - Adar(u, v) = \sum_{z \in \Gamma(u) \cap \Gamma(v)} \frac{1}{\log |\Gamma(z)|}. \tag{2.3}$$

Based on the past results of existing literature on link prediction, Adamic-Adar usually works better than the previous two metrics.

Path-Based Metrics

Shortest Path Distance. The basic idea of this method is that if two nodes' distance (by shortest path) in a social network is short, then it is likely that a link would be created between the two nodes. However, according to the theory of six degrees of separation, everyone and everything is six or fewer steps away [107]. Thus, this feature sometimes does not work that well.

Katz. Leo Katz proposed this metric as an extension of the shortest path distance. It works much better for link prediction. It directly sums over all the paths that exist between a pair of nodes u and v. To penalize the contribution of longer paths in the similarity computation it exponentially damps the contribution of a path by a factor of β_l, where l is the path length. The exact equation to compute the Katz score is as below:

$$Katz(u, v) = \sum_{l=1}^{\infty} \beta_l \cdot |paths_{u,v}^{(l)}|, \tag{2.4}$$

where $paths\langle l\rangle$ is a set of paths of length l from u to v. The parameter $\beta(\leq 1)$ can be used to regularize this feature. A small β indicates that we only consider shorter paths. In an extreme case, this metric would behave like node neighborhood-based metrics. One problem with this feature is that it is computationally expensive. It can be shown that the Katz score between all the pairs of nodes can be computed by finding $(I - \beta A)^{-1} - I$, where A is the adjacency matrix and I is an identity matrix of proper size. This task has roughly cubic complexity which could be infeasible for large social networks.

Hitting Time. The concept of hitting time comes from random walks on a graph. For two nodes, u and v in a graph, the hitting time, $H_{u,v}$ denotes the expected number of steps required for a random walk starting at u to reach v. Shorter hitting time denotes that the nodes are similar to each other, so they have a higher chance of linking in the future. Since this metric is not symmetric, for undirected graphs the commute time, $C_{u,v} = H_{u,v} + H_{v,u}$, can be used. The benefit of this metric is that it is easy to compute by performing some trial random walks. On the downside, its value can have high variance; hence, performance of link prediction by this metric could be not ideal. For instance, the hitting time between u and v can be affected by a node w, which is far away from u and v; for instance, if w has high stationary probability, then it could be hard for a random walk to escape from the neighborhood of w. To protect against this problem we can use random walks with restart, where we periodically reset the random walk by returning to u with a fixed probability α in each step. Due to the scale-free nature of a social network some of the nodes may have very high stationary probability (π) in a random walk; to safeguard against it, the hitting time can be normalized by multiplying it with the stationary probability of the respective node, as shown below:

$$Normalized - Hitting - Time(u, v) = H_{u,v} \cdot \pi_v + H_{v,u} \cdot \pi_u. \tag{2.5}$$

Rooted Pagerank. Pagerank value can also be used as a metric for link prediction. However, since Pagerank itself quantifies the importance of a single vertex, it requires to be modified so that it can estimate a similarity between a pair of vertices u and v. The original definition of Pagerank denotes the importance of a vertex under two assumptions: for some fixed probability α, a surfer at a web-page jumps to a random web-page with probability α and follows a linked hyperlink with probability $1 - \alpha$. Under this random walk, the importance of a web-page v is the expected sum of the importance of all the web-pages u that link to v. In random walk terminology, one can replace the term importance by the term stationary distribution. For link prediction, the random walk assumption of the original Pagerank can be altered as below: similarity score between two nodes u and v can be measured as the stationary probability of v in a random walk that returns to u with probability $1 - \alpha$ in each step, moving to a random neighbor with probability α. This metric is asymmetric and can be made symmetric by summing with the counterpart where the role of u and v are reversed, which is also named as rooted Pagerank. The rooted Pagerank between all node pairs (represented as RPR) can be derived as follows. Let D be a diagonal degree matrix with $D[i, i] = \sum_j A[i, j]$. Let, $N = D^{-1}A$ be the adjacency matrix with row sums normalized to 1. Then,

$$RPR = (1 - \alpha) \cdot (I - \alpha N)^{-1}. \tag{2.6}$$

Metrics Based on Node and Edge Attributes

Vertex and edge attributes play an important role for link prediction. Note that in a social network the links are directly motivated by the utility of the individual representing the nodes and the utility is a function of node and edge attributes. Many studies showed that node or edge attributes as proximity features can significantly increase the performance of link prediction tasks. For example, Hasan et al. [64] showed that, for link prediction in a co-authorship social network, attributes such as the degree of overlap among the research keywords used by a pair of authors is the top ranked attribute for some datasets. Here the node attribute is the research keyword set and the assumption is that a pair of authors are close (in the sense of a social network) to each other, if their research work evolves around a larger set of common keywords. Similarly, the Katz metric computed the similarity between two web-pages by the degree to which they have a larger set of common words where the words in the web-page are the vertex attributes. The advantage of such a metric set is that it is generally cheap to compute. On the down-side, the metrics are very tightly tied with the domain, so, it requires good domain knowledge to identify them. Below, we will provide a generic approach to show how these features can be incorporated into a link prediction task.

Node Feature Aggregation. Once we identify an attribute a of a node in a social network, we need to devise some meaningful aggregation function f. To compute the similarity value between nodes u and v, f accepts the corresponding attribute values of these nodes to produce a similarity score. The choice of function entirely depends on the type of the attribute. In the following we show two examples where we aggregated some local metric of a node.

- *Preferential Attachment Score.* In general, a node connects to other nodes in the network based on the probability of their degree. So, if we consider the neighborhood size as feature value, then multiplication can be an aggregation function, which is named as preferential attachment score:

$$Preferential - Attachment - Score(u, v) = \Gamma(u) \cdot \Gamma(v). \tag{2.7}$$

- *Clustering Coefficient Score.* Clustering coefficient of a vertex u is denoted as below:

$$Clustering - Coefficient(u) = \frac{3 \times \#triangles\ adjacent\ to\ v}{\#possible\ triples\ adjacent\ to\ v}. \tag{2.8}$$

To compute a score for link prediction between the vertex u and v, one can sum or multiply the clustering coefficient score of u and v.

Generic SimRank. Jeh and Widom suggested a generic metric called SimRank which recursively captures that "two objects are similar if they are similar to two similar objects." The SimRank score is the fixed point of the following recursive equation:

$$SimRank(u, v) = \begin{cases} 1 & if\ u = v \\ \gamma \cdot \frac{\sum_{a \in \Gamma(u)} \sum_{b \in \Gamma(v)} SimRank(a,b)}{|\Gamma(u)| \cdot |\Gamma(v)|} & otherwise \end{cases}. \tag{2.9}$$

Latent Feature-based Metrics

In many link prediction problems, while the feature vectors representing data objects are high-dimensional, the number of latent features actually effective for predicting links is assumed to be relatively small. Therefore, the accuracy of link prediction can be improved by identifying and working in a low-dimensional latent feature space. In supervised linear dimension-reduction methods, a linear projection $\mathbf{W}$ from the original D-dimensional feature space to a $d(< D)$-dimensional latent feature space is learned from training data consisting of data objects known to have or not to have links between them. The learning process seeks the linear projection $\mathbf{W}$ that makes the distance in the mapped space,

$$\| \mathbf{W_x} - \mathbf{W_y} \|, \tag{2.10}$$

as small as possible, where $\mathbf{x}$ and $\mathbf{y}$ are two nodes known to have a link between them. After the learning process is completed, two data objects with an unknown link status are mapped to the latent space by using $\mathbf{W}$. If the mapped images of the two data objects are sufficiently close to each other, they are considered to have a link between them.

Assume that we have N training data objects, $\mathbf{x}_1, \ldots, \mathbf{x}_N$, and that each data object $\mathbf{x}_i$ is represented in a D dimensional feature vector [115]. One can use locality preserving projections to find the optimal linear projection matrix $\mathbf{W}^*$ by solving the following optimization problem:

$$\mathbf{W}^* = \arg\min_{\mathbf{W}} \sum_{i,j} A_{ij} \| \mathbf{W_{x_i}} - \mathbf{W_{x_j}} \|_2^2, \tag{2.11}$$

where $\| \cdot \|_2$ is the Euclidean norm (2-norm), and $\mathbf{A} = \{A_{ij}\}$ is the adjacency matrix defined by

$$A_{ij} = \begin{cases} 1 & if\ \mathbf{x}_i\ and\ \mathbf{x}_j\ have\ a\ link \\ 0 & otherwise. \end{cases} \tag{2.12}$$

Challenge for Link Prediction as Classification

The main challenge in similarity metric-based supervised link prediction is extreme class skewness. The number of possible links is quadratic in the number of vertices in a social network, however the number of actual links (the edges in the graph) added to the graph is only a tiny fraction of this number. This results in large class skewness, causing training and inference to become difficult tasks [64].

The problem of class skewness in supervised learning is well known in machine learning. The poor performance of a learning algorithm in this case results from both the variance in the models estimates and the imbalance in the class distribution. Even if a low proportion of negative instances have the predictor value similar to the positive instances, the model will end up with a large raw number of false positives.

To cope with class skew, existing research suggests several different approaches. These methods include the altering of the training sample by up-sampling or down-sampling, altering the learning method by making the process active or cost-sensitive, and also more generally by treating the classifier score with different thresholds. In general, learning from imbalanced datasets is a very important research consideration and Weiss [161] has a good discussion of various techniques to solve this.

2.2.2   MATRIX FACTORIZATION

We briefly introduce two series of matrix factorization based methods, which are to predict undirected and directed links, respectively.

Predicting Undirected Links using Spectral Graph Transformations

Approaching the problem of link prediction algebraically, we can consider a graph adjacency matrix $\mathbf{A}$, and look for a function $F(\mathbf{A})$ returning a matrix of the same size whose entries can be used for prediction. Kunegis et al.'s approach consists of computing a matrix decomposition $\mathbf{A} = \mathbf{U}\mathbf{D}\mathbf{V}^T$ and considering functions of the form $F(\mathbf{A}) = \mathbf{U} F(\mathbf{D})\mathbf{V}^T$, where $F(\mathbf{D})$ applies a function on reals to each element of the graph spectrum $\mathbf{D}$ separately [86]. The authors show that a certain number of common links and edge weight prediction algorithms can be mapped to this form. As a result, the method provides a mechanism for estimating any parameters of such link prediction algorithms. Analogously, they also consider a network's Laplacian matrix as the basis for link prediction.

Let $\mathbf{A} \in \{0, 1\}^{n \times n}$ be the adjacency matrix of a simple, undirected, unweighted, and connected graph on n nodes, and $F(\mathbf{A})$ a function that maps $\mathbf{A}$ to a matrix of the same dimension.

The following subsections describe link prediction functions $F(\mathbf{A})$ that result in matrices of the same dimension as $\mathbf{A}$ and whose entries can be used for link prediction. Most of these methods result in a positive-semidefinite matrix, and can be qualified as graph kernels. The letter α will be used to denote parameters of these functions.

Functions of the Adjacency Matrix. Let $\mathbf{D} \in \mathbb{R}^{n \times n}$ be the diagonal degree matrix with $\mathbf{D}_{ii} = \sum_j \mathbf{A}_{ij}$. Then $\mathcal{A} = \mathbf{D}^{-1/2}\mathbf{A}\mathbf{D}^{-1/2}$ is the normalized adjacency matrix. Transformations of the adjacency matrices $\mathbf{A}$ and $\mathcal{A}$ give rise to the exponential and von Neumann graph kernels:

$$F_{EXP}(\mathbf{A}) = \exp(\alpha \mathbf{A}) \tag{2.13}$$

$$F_{EXP}(\mathcal{A}) = \exp(\alpha\mathcal{A}) \tag{2.14}$$

$$F_{NEU}(\mathbf{A}) = (\mathbf{I} - \alpha\mathbf{A})^{-1} \tag{2.15}$$

$$F_{NEU}(\mathcal{A}) = (\mathbf{I} - \alpha\mathcal{A})^{-1}, \tag{2.16}$$

where $\alpha < 1$ is a positive parameter. The constraint $\alpha < 1$ is required by the von Neumann kernels.

Laplacian Kernels. $\mathbf{L} = \mathbf{D}\text{-}\mathbf{A}$ is the combinatorial Laplacian of the graph, and $\mathcal{L} = \mathbf{I}\text{-}\mathcal{A} = \mathbf{D}^{-1/2}\mathbf{L}\mathbf{D}^{-1/2}$ is the normalized Laplacian. The Laplacian matrices are singular and positive-semidefinite. Their Moore-Penrose pseudoinverse is called the commute time or resistance distance kernel. The combinatorial Laplacian matrix is also known as the Kirchhoff matrix, due to its connection to electrical resistance networks:

$$F_{COM}(\mathbf{L}) = \mathbf{L}^{+} \tag{2.17}$$

$$F_{COM}(\mathcal{L}) = \mathcal{L}^{+}. \tag{2.18}$$

By regularization, the regularized Laplacian kernels are:

$$F_{COMR}(\mathbf{L}) = (\mathbf{I} + \alpha\mathbf{L})^{-1} \tag{2.19}$$

$$F_{COMR}(\mathcal{L}) = (\mathbf{I} + \alpha\mathcal{L})^{-1}. \tag{2.20}$$

As a special case, the non-normalized regularized Laplacian kernel is called the random forest kernel for $\alpha = 1$. The normalized regularized Laplacian is equivalent to the normalized von Neumann kernel by noting that $(1 + \alpha\mathcal{L})\text{-}1 = (1 + \alpha)(\mathbf{I}\text{-}\alpha\mathcal{A})\text{-}1$.

The heat diffusion kernel is defined as

$$F_{HEAT}(\mathbf{L}) = \exp(-\alpha\mathbf{L}) \tag{2.21}$$

$$F_{HEAT}(\mathcal{L}) = \exp(-\alpha\mathcal{L}). \tag{2.22}$$

The normalized heat diffusion kernel is equivalent to the normalized exponential kernel: $\exp(-\alpha\mathcal{L}) = e^{-\alpha} \exp(\alpha\mathcal{A})$.

Rank Reduction. Using the eigenvalue decomposition $\mathbf{A} = \mathbf{U}\mathbf{\Lambda}\mathbf{U}^{T}$, a rank-$k$ approximation of $\mathbf{A}$, $\mathbf{L}$, $\mathcal{A}$, and $\mathcal{L}$ is given by a truncation leaving only k eigenvalues and eigenvectors in $\mathbf{\Lambda}$ and $\mathbf{U}$:

$$F_{(k)}(\mathbf{A}) = \mathbf{U}_{(k)}\mathbf{\Lambda}_{(k)}\mathbf{U}_{(k)}^{T}. \tag{2.23}$$

For $\mathbf{A}$ and $\mathcal{A}$, the biggest eigenvalues are used while the smallest eigenvalues are used for the Laplacian matrices. $F_{(k)}(\mathbf{A})$ can be used for prediction itself, or serve as the basis for any of the graph kernels. In practice, only rank-reduced versions of graph kernels can be computed for large networks.

Path Counting. One can exploit the fact that powers $\mathbf{A}^{n}$ of the adjacency matrix of an unweighted graph contain the number of paths of length n connecting all node pairs. On the basis that nodes connected by many paths should be considered nearer to each other than nodes connected by few paths, a weighted mean of powers of $\mathbf{A}$ can be computed as a link prediction function

$$F_P(\mathbf{A}) = \sum_{i=0}^{d} \alpha_i \mathbf{A}^i. \tag{2.24}$$

The result is a matrix polynomial of degree d. The coefficients α_i should be decreasing to reflect the assumption that links are more likely to arise between nodes that are connected by short paths than nodes connected by long paths. Thus, such a function takes both path lengths and path counts into account.

The exponential and von Neumann kernels can be expressed as infinite series of matrix powers:

$$\exp(-\alpha \mathbf{A}) = \sum_{i=0}^{\infty} \frac{\alpha^i}{i!} \mathbf{A}^i \tag{2.25}$$

$$(\mathbf{I} - \alpha \mathbf{A})^{-1} = \sum_{i=0}^{\infty} \alpha^i \mathbf{A}^i. \tag{2.26}$$

Generalization. All these link prediction methods can be written as $\mathbf{F} = F(\mathbf{X})$, where $\mathbf{X}$ is one of $\{\mathbf{A}, \mathcal{A}, \mathbf{L}, \mathcal{L}\}$ and F is either a matrix polynomial, matrix (pseudo) inversion, the matrix exponential or a function derived piecewise linearly from one of these. Such functions F have the property that for a symmetric matrix $\mathbf{A} = \mathbf{U}\mathbf{\Lambda}\mathbf{U}^T$, they can be written as $F(\mathbf{A}) = \mathbf{U}F(\mathbf{\Lambda})\mathbf{U}^T$, where $F(\mathbf{\Lambda})$ applies the corresponding function on reals to each eigenvalue separately. In other words, these link prediction methods result in prediction matrices that are simultaneously diagonalizable with the known adjacency matrix. Such functions are called spectral transformations and can be written as $F \in \mathcal{S}$.

Given a graph G, the target is to find a spectral transformation F that performs well at link prediction for this particular graph. To that end, the edge set of G is devided into a training set and a test set, and then the purpose is to look for an F that maps the training set to the test set with minimal error.

Formally, let $\mathbf{A}$ and $\mathbf{B}$ be the adjacency matrices of the training and test set, respectively. $\mathbf{A}$ is the source matrix and $\mathbf{B}$ is the target matrix. The solution to the following optimization problem gives the optimal spectral transformation for the task of predicting the edges in the test set.

Problem 2.1 Let $\mathbf{A}$ and $\mathbf{B}$ be two adjacency matrices over the same vertex set. A spectral transformation that maps $\mathbf{A}$ and $\mathbf{B}$ with minimal error is given by the solution to

$$\min_{F} \| F(\mathbf{A}) - \mathbf{B} \|_F \quad s.t. \quad F \in \mathcal{S} \tag{2.27}$$

where $\| \cdot \|_F$ denotes the Frobenius norm.

Table 2.2: Summary of network datasets used in the experiments [86]

Name	#Vertices	#Edges	Weights	k	Description
DBLP	12,563	49,779	$\{1\}$	126	Citation graph
Hep-th	27,766	352,807	$\{1\}$	54	Citation graph
Advogato	7,385	57,627	$\{0.6, 0.8, 1.0\}$	192	Trust network
Slashdot	71,523	488,440	$\{1, +1\}$	24	Friend/foe network
Epinions	131,828	841,372	$\{1, +1\}$	14	Trust/distrust network
WWW	325,729	1,497,135	$\{1\}$	49	Hyperlink graph
WT10G	1,601,787	8,063,026	$\{1\}$	49	Hyperlink graph

Problem 2.1 can be solved by computing the eigenvalue decomposition $\mathbf{A} = \mathbf{U\Lambda U}^T$ and using the fact that the Frobenius norm is invariant under multiplication by an orthogonal matrix

$$\|F(\mathbf{A}) - \mathbf{B}\|_F = \|\mathbf{U}F(\mathbf{\Lambda})\mathbf{U}^T - \mathbf{B}\|_F = \|F(\mathbf{\Lambda}) - \mathbf{U}^T\mathbf{B}\mathbf{U}\|_F. \tag{2.28}$$

The Frobenius norm in (2.28) can be decomposed into the sum of squares of off-diagonal entries of $F(\mathbf{\Lambda}) - \mathbf{U}^T\mathbf{B}\mathbf{U}$, which is independent of F, and into the sum of squares of its diagonal entries. This leads to the following least-squares problem equivalent to Problem 2.1.

Problem 2.2   If $\mathbf{U\Lambda U}^T$ is the eigenvalue decomposition of $\mathbf{A}$, then the solution to Problem 2.1 is given by $F(\mathbf{\Lambda})_{ii} = f(\mathbf{\Lambda}_{ii})$, where $f(x)$ is the solution to the following minimization problem.

$$\min_f \sum_i (f(\mathbf{\Lambda}_{ii}) - \mathbf{U}_{\cdot i}^T\mathbf{B}\mathbf{U}_{\cdot i})^2. \tag{2.29}$$

This problem is a one-dimensional least-squares curve fitting problem of size n. Since each function $F(\mathbf{A})$ corresponds to a function $f(x)$, one can choose a link prediction function F and learn its parameters by inspecting the corresponding curve fitting problem.

Results. The inspected network datasets are summarized in Table 2.2. DBLP[2] is a citation graph. Hep-th is the citation graph of Arxiv's high energy physics/theory section. Advogato is a trust network with three levels of trust. Slashdot is a social network where users tag each other as friends and foes. Epinions is an opinion site where users can agree or disagree with each other. WWW and WT10G are hyperlink network datasets extracted from a subset of the World Wide Web.

As a measure of performance for the link prediction methods, the Pearson correlation coefficient between the predicted and known ratings in the test set is computed. The correlation represents a trade-off between the root mean squared error to minimize and more dataset-dependent measures such as precision and recall, which cannot be applied to all datasets. The results of the evaluation are shown in Table 2.3.

[2]http://dblp.uni-trier.de/

Table 2.3: The results of the experimental evaluation. For each dataset, the source/target matrices, the curve fitting model and the link prediction method that perform best are given [86]

Dataset	Best transformation	Best fitting curve	Best graph kernel	Correlation
DBLP	$L \rightarrow B$	Polynomial	Sum of powers	0.563
Hep-th	$\mathcal{A} \rightarrow B$	Exponential	Heat diffusion	0.453
Advogato	$\mathcal{L} \rightarrow B$	Rational	Commute time	0.554
Slashdot	$A \rightarrow B$	Nonnegative odd polynomial	Sum of powers	0.263
Epinions	$A \rightarrow A + B$	Nonnegative odd polynomial	Sum of powers	0.354
WWW	$L \rightarrow A + B$	Polynomial	Sum of powers	0.739
WT10G	$A \rightarrow B$	Linear function	Rank reduction	0.293

2.3   INFERRING SOCIAL TIES

The objective of inferring social ties is to effectively infer the type of social relationships between two users. More precisely, we first define the output of our problem, namely *relationship semantics*.

Specifically, given users' behavior history and interactions between users, can we estimate how likely they are to be family members or colleagues? One challenge is how to design a unified model so that it can be easily applied to different domains (or different networks)? There exist a few related studies. For example, Diehl et al. [38] tried to identify the relationships by learning a ranking function. Wang et al. [155] proposed an unsupervised algorithm for mining the advisor-advisee relationships from the publication network. However, Dieh et al. [38] only considers the communication archive, while Wang et al. [155] is a domain-specific unsupervised algorithm. Both algorithms are not easy to extend to other domains.

Another challenge is that online social networks are becoming more and more complex and dynamic. Even the best performance achieved by the state-of-the-art algorithms is still under 90%. The result is unsatisfactory and invariably contains a number of errors. A promising solution is to design an interactive interface to allow users to provide feedbacks on the inferring results. However, we should be aware that the interactive process might be tedious, error-prone, and time-consuming. For example, for inferring advisor-advisee relationships from the co-author network, an author may have hundreds of co-authors.[3] The user may soon become tired, if she/he is asked to carefully go through all her/his relationships to validate the inferred results. Ideally, an algorithm should be able to actively select only a few potentially wrong relationships to query the user, instead of passively waiting for user feedbacks. The problem is referred to as *actively learning to infer social ties*.

To illustrate the problem, Figure 2.2 gives an example of actively inferred ties in a mobile communication network. The left figure gives the input of our problem: a mobile social network, which consists of users, calls made and messages sent between users, and users' attribute information such as location. The objective is to classify the type of social relationships in the network. The middle figure shows the result of the proposed PLP-FGM model, a semi-supervised learning model. The blue solid lines stand for friend relationship between users and the green dash lines indicate colleagues. The probability associated with each relationship represents how confident the learning model is in the inferred type of the relationship. Further, an active learning algorithm selects a uncertain relationship (associated with a question mark) to query the user. Once the user gives the answer, the learning model propagates the correction in the social network and further corrects other relationships.

Therefore, the fundamental problem is how to design a flexible model for effectively and efficiently learning to infer social ties in different networks. This problem is non-trivial and poses a set of unique challenges. First, what are the underlying factors that form a specific type of social relationship? Second, the input social network is partially labeled. We may have some labeled relationships, but most of the relationships are unknown. To learn a high-quality predictive model,

---

[3]An example can be found on http://arnetminer.org/person/jiawei-han-745329.html

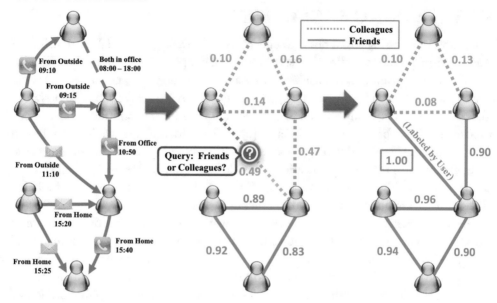

Figure 2.2: An example of learning to infer social ties in a mobile communication network. The left figure is the input of our problem and the middle figure shows the output (the inferred relationships) of the problem. The relationship associated with the question mark indicates a relationship selected by an active learning algorithm to query the user. The right figure is the improved result with the user's feedback (via active learning).

we should not only consider the knowledge provided by the labeled relationships, but also leverage the unlabeled network information. Third, how does one make optimal use of user interaction? The selection should consider both the uncertainty and the network structure information. Finally, real social networks are getting bigger with thousands, even millions, of nodes. It is important to develop a method that can scale well to real large networks.

In the following, we formally formulate the problem of inferring a social relationship in large networks, and propose a partially labeled pairwise factor graph model (PLP-FGM). To make optimal use of user interactions, two strategies—an influence maximization-based strategy and a belief maximization-based strategy—were devised to actively select potentially wrong but most useful relationships to query the user. We further extend the model by incorporating social theories into the semi-supervised learning framework. In this way, the model is able to support transferring supervised information from a source network to help infer social ties in a heterogeneous target network.

2.3.1   PROBLEM FORMULATION

The input is still a social network $G = (V, E)$. The output of the problem here are *relationship semantics*.

Definition 2.3   Relationship semantics: Relationship semantics is a triple (e_{ij}, r_{ij}, p_{ij}), where $e_{ij} \in E$ is a social relationship; $r_{ij} \in \mathcal{Y}$ is a label associated with the relationship; $\mathcal{Y}$ is the set of all the labels; p_{ij} is the probability (confidence) obtained by an algorithm for inferring relationship type.

Social relationships might be undirected in some networks (e.g., the friendship discovered from the mobile communication network) or directed in other networks (e.g., the advisor-advisee relationship in the publication network). To be consistent, we define all social relationships as directed relationships. In addition, relationships may be static (e.g., the family-member relationship) or dynamic over time (e.g., colleague relationship). In this book, we focus on static relationships, and leave the dynamic case to our future work.

To infer relationship semantics, we could consider different factors such as user-specific information, link-specific information, and global constraints (cf. Table 2.6 for examples). For example, to discover advisor-advisee relationships from a publication network, we can consider how many papers were co-authored by two authors; how many papers in total an author has published; and when the first paper was published by each author. Besides, there may exist some labeled relationships. Formally, we can define the input of our problem, a partially labeled network.

Definition 2.4   Partially labeled network: A partially labeled network is an augmented social network denoted as $G = (V, E^L, E^U, R^L, \mathbf{W})$, where E^L is a set of labeled relationships and E^U is a set of unlabeled relationships with $E^L \cup E^U = E$; R^L is a set of labels corresponding to the relationships in E^L; $\mathbf{W}$ is an attribute matrix associated with users in V where each row corresponds to a user, each column an attribute, and an element w_{ij} the value of the jth attribute of user v_i.

Based on the above concepts, we can define the problem of inferring social relationships. Given a partially labeled network, the goal is to detect the types (labels) of all unknown relationships in the network. More precisely, we have the following.

Problem 2.5   Social relationship mining. Given a partially labeled network $G = (V, E^L, E^U, R^L, \mathbf{W})$, the objective is to learn a predictive function

$$f : G = (V, E^L, E^U, R^L, \mathbf{W}) \to R.$$

Another important question is how we can learn the mapping function f effectively. In many situations, labeled data is limited and expensive. The problem is, can we design a strategy to *actively* learn the model with minimal labeling cost? Formally, we have the following.

Problem 2.6   Active relationship mining. Given a partially labeled network $G = (V, E^L, E^U, R^L, \mathbf{W})$, and a labeling budget b (number of user interactions). Our objective is to select a subset of unknown relationships $A \subset E^U$ within the constraint of b to label, so that the performance of predictive function f can be maximally improved.

Therefore, the problem is how to find a function f that can leverage both the labeled relationships and the unlabeled relationships to infer the unknown relationships.

2.3.2   UNSUPERVISED LEARNING TO INFER SOCIAL TIES

Acquiring sufficient labeled relationships is always expensive. Let us begin with the unsupervised learning method for inferring the type of social relationships without labeled data. Such a method is usually task-oriented. For example, Wang et al. [155] proposed a two-stage framework, referred to as TPFG, for inferring advisor-advisee relationships in the co-author network. The main idea is to leverage a time-constrained probabilistic factor graph model to decompose the joint probability of the unknown advisor of every author. The time-related information associated to the hidden social role is captured via factor functions, which form the basic components of the factor graph model. By maximizing the joint probability of the factor graph one can infer the relationship and compute a ranking score for each relationship on the candidate graph.

More specifically, at the first stage of the framework, common sense is defined as recognizing interesting semantic relationships. Here the authors try to make a few general assumptions based on common knowledge about advisor-advisee relationships.

- At each time t during the publication history of an author x, x is either being advised or not being advised. Once x starts to advise another author, it will not be advised again.

- Another assumption indicates that, for a given pair of advisor and advisee, the advisor always has a longer publication history than the advisee.

Based on the two assumptions, the framework processes the task in the following two stages.

Stage 1: Preprocessing. The purpose of preprocessing is to generate the candidate graph H' and reduce the search space while keeping the real advisor not excluded from the candidate pool in most cases. First, one needs to generate according to the co-author information a homogeneous author network G' by processing the papers in the network one by one. For each paper p_i, we can construct an edge between every pair of its authors.

Then a filtering process is performed to remove unlikely relations of advisor-advisee. For each edge e_{ij} on G', a_i and a_j has collaboration. To decide whether a_j is a_i's potential advisor, the following conditions are checked. First, the second assumption is checked. Only if a_j

started to publish earlier than a_i, is the possibility considered. Second, some heuristic rules are applied, which are based on the prior intuitive knowledge about advisor-advisee relations. For more detailed definitions of those rules, please refer to Wang et al. [155].

Stage 2: The factor graph model. From the candidate graph H' we know the potential advisors of each author and the likelihood based on local information. By modeling the network as a whole, we can incorporate both structural information and temporal constraint and better analyze the relationship among individual links.

By learning the factor graph model, we can find a configuration of the latent variables for each node in the candidate graph H' that maximize the objective function. For learning the model, one can consider the sum-product and the junction tree algorithms [155].

Results. To evaluate the unsupervised method, we tried to collect an academic citation database (ArnetMiner [148][4]), and the labeled advisor-advisee relationships from several online resources, such as the Mathematics Genealogy project and AI Genealogy project. The database consists of 654,628 authors and 1,076,946 publications from 1970–2008. By applying the proposed TPFG model to the publication dataset, we could achieve a performance of 81–85% (in terms of F1-Measure).

2.3.3   SUPERVISED LEARNING TO INFER SOCIAL TIES

We now introduce how to leverage supervised learning for inferring social ties. Basically, we have three basic intuitions. First, the user-specific or link-specific attributes will contain implicit information about the relationships. For example, two users who make a number of calls in working hours might be colleagues, while two users who frequently contact with each other in the evening are more likely to be family members or intimate friends. Second, relationships among different users may have a correlation. For example, in the mobile network, if user v_i makes a call to user v_j immediately after calling user v_k, then user v_i may have a similar relationship (family member or colleague) with user v_j and user v_k. Third, we also need to consider some global constraints such as common knowledge or user-specific constraints.

Based on the intuitions above, we propose a Partially-Labeled Pairwise Factor Graph Model (PLP-FGM). This allows us to take all of the factors mentioned above into account to better infer the social relationships. Typically, there are two ways to model the social tie inferring problem. The first way is to model each user as a node and for each node to estimate the probability distribution of different relationships. The resultant graphical model thus consists of N variable nodes. Each node contains a $d \times |\mathcal{Y}|$ matrix to represent the probability distributions of different relationships between the user and her/his neighbors, where d is the number of neighbors of the node. This model is intuitive, but it suffers from some limitations. For example, it is difficult to model the correlations between two relationships, and its computational complexity is high. An alternative way is to model each relationship as a node in the graphical model and the relationship

---

[4]http://aminer.org

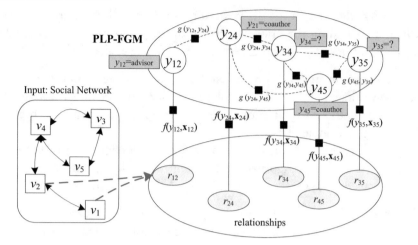

Figure 2.3: Graphical representation of the PLP-FGM model

mining task becomes how to predict the semantic label for each relationship node in the model. This model contains M nodes ($2M$ when the input social network is undirected). This model is able to incorporate different correlations between relationships such as the above intuitions.

We propose a Partially-Labeled Pairwise Factor Graph Model (PLP-FGM). Figure 2.3 shows the graphical representation of the PLP-FGM. Each relationship (v_{i_1}, v_{i_2}) or $e_{i_1 i_2}$ in the partially labeled network G is mapped to a *relationship node* r_i in PLP-FGM. We denote the set of relationship nodes as $Y = \{y_1, y_2, \ldots, y_M\}$. The relationships in G are partially labeled, thus all nodes in PLP-FGM can be divided into two subsets Y^L and Y^U, corresponding to the labeled and unlabeled relationships, respectively. For each relationship node $y_i = (v_{i_1}, v_{i_2}, r_{i_1 i_2})$, we combine the attributes $\{\mathbf{w}_{i_1}, \mathbf{w}_{i_2}\}$ into a *relationship attribute vector* $\mathbf{x}_i$.

Now we explain the PLP-FGM in detail. The relationships in the input are modeled by relationship nodes in PLP-FGM. Corresponding to the three intuitions, we define the following three factors.

- *Attribute factor*: $f(y_i, \mathbf{x}_i)$ represents the posterior probability of the relationship y_i given the attribute vector $\mathbf{x}_i$.

- *Correlation factor*: $g(y_i, G(y_i))$ denotes the correlation between the relationships, where $G(y_i)$ is the set of correlated relationships to y_i.

- *Constraint factor*: $h(y_i, H(y_i))$ reflects the constraints between relationships, where $H(y_i)$ is the set of relationships constrained on y_i.

Given a partially labeled network $G = (V, E^L, E^U, R^L, \mathbf{W})$, we can define the joint distribution over Y as

$$p(Y|G) = \prod_i f(y_i, \mathbf{x}_i) g(y_i, G(y_i)) h(y_i, H(y_i)). \tag{2.30}$$

The three factors can be instantiated in different ways. In this book, we use exponential-linear functions. In particular, we define the attribute factor as

$$f(y_i, \mathbf{x}_i) = \frac{1}{Z_\lambda} \exp\{\boldsymbol{\lambda}^T \boldsymbol{\Phi}(y_i, \mathbf{x}_i)\}, \tag{2.31}$$

where $\boldsymbol{\lambda}$ is a weighting vector and $\boldsymbol{\Phi}$ is a vector of feature functions. Similarly, we define the correlation factor and constraint factor as

$$g(y_i, G(y_i)) = \frac{1}{Z_\alpha} \exp\{ \sum_{y_j \in G(y_i)} \boldsymbol{\alpha}^T \mathbf{g}(y_i, y_j)\} \tag{2.32}$$

$$h(y_i, H(y_i)) = \frac{1}{Z_\beta} \exp\{ \sum_{y_j \in H(y_i)} \boldsymbol{\beta}^T \mathbf{h}(y_i, y_j)\}, \tag{2.33}$$

where $\mathbf{g}$ and $\mathbf{h}$ can be defined as a vector of indicator functions. This feature definition was often used in a graphical models such as Markov Random Fields [63] or Conditional Random Fields [87].

Model Learning Learning PLP-FGM is to estimate a parameter configuration $\boldsymbol{\theta} = (\boldsymbol{\lambda}, \boldsymbol{\alpha}, \boldsymbol{\beta})$, so that the log-likelihood of observation information (labeled relationships) are maximized. For presentation simplicity, we concatenate all factor functions for a relationship node y_i as $\mathbf{s}(y_i) = (\boldsymbol{\Phi}(y_i, \mathbf{x}_i)^T, \sum_{y_j} \mathbf{g}(y_i, y_j)^T, \sum_{y_j} \mathbf{h}(y_i, y_j)^T)^T$. The joint probability defined in (Eq. (2.30)) can be rewritten as

$$p(Y|G) = \frac{1}{Z} \prod_i \exp\{\boldsymbol{\theta}^T \mathbf{s}(y_i)\} = \frac{1}{Z} \exp\{\boldsymbol{\theta}^T \sum_i \mathbf{s}(y_i)\} = \frac{1}{Z} \exp\{\boldsymbol{\theta}^T \mathbf{S}\}, \tag{2.34}$$

where $Z = Z_\lambda Z_\alpha Z_\beta$ is a normalization factor (also-called partition function), $\mathbf{S}$ is the aggregation of factor functions over all relationship nodes, i.e., $\mathbf{S} = \sum_i \mathbf{s}(y_i)$.

One challenge for learning the PLP-FGM model is that the input data is partially labeled. To calculate the partition function Z, one needs to sum up the likelihood of possible states for all nodes including unlabeled nodes. To deal with this, we use the labeled data to infer unknown labels. Here, $Y|Y^L$ denotes a labeling configuration Y inferred from the known labels. Thus, we can define the following log-likelihood objective function $\mathcal{O}(\boldsymbol{\theta})$:

Input: learning rate η
Output: learned parameters $\boldsymbol{\theta}$

Initialize $\boldsymbol{\theta}$;
repeat

    Calculate $\mathbb{E}_{p_{\theta}(Y|Y^L,G)}\mathbf{S}$ using LBP ;
    Calculate $\mathbb{E}_{p_{\theta}(Y|G)}\mathbf{S}$ using LBP ;
    Calculate the gradient of θ according to Eq. (2.36):

$$\nabla_{\boldsymbol{\theta}} = \mathbb{E}_{p_{\theta}(Y|Y^L,G)}\mathbf{S} - \mathbb{E}_{p_{\theta}(Y|G)}\mathbf{S}$$

    Update parameter θ with the learning rate η:

$$\boldsymbol{\theta}_{\text{new}} = \boldsymbol{\theta}_{\text{old}} - \eta \cdot \nabla_{\boldsymbol{\theta}}$$

until *Convergence*;

Algorithm 1: Learning PLP-FGM.

$$
\begin{aligned}
\mathcal{O}(\boldsymbol{\theta}) &= \log p(Y^L|G) = \log \sum_{Y|Y^L} \frac{1}{Z} \exp\{\boldsymbol{\theta}^T \mathbf{S}\} \\
&= \log \sum_{Y|Y^L} \exp\{\boldsymbol{\theta}^T \mathbf{S}\} - \log Z \\
&= \log \sum_{Y|Y^L} \exp\{\boldsymbol{\theta}^T \mathbf{S}\} - \log \sum_{Y} \exp\{\boldsymbol{\theta}^T \mathbf{S}\}.
\end{aligned}
\tag{2.35}
$$

To solve the objective function, we consider a gradient decent method (or a Newton-Raphson method). Specifically, we first calculate the gradient for each parameter θ:

$$
\begin{aligned}
\frac{\partial \mathcal{O}(\boldsymbol{\theta})}{\partial \boldsymbol{\theta}} &= \frac{\partial \left(\log \sum_{Y|Y^L} \exp \boldsymbol{\theta}^T \mathbf{S} - \log \sum_{Y} \exp \boldsymbol{\theta}^T \mathbf{S} \right)}{\partial \boldsymbol{\theta}} \\
&= \frac{\sum_{Y|Y^L} \exp \boldsymbol{\theta}^T \mathbf{S} \cdot \mathbf{S}}{\sum_{Y|Y^L} \exp \boldsymbol{\theta}^T \mathbf{S}} - \frac{\sum_{Y} \exp \boldsymbol{\theta}^T \mathbf{S} \cdot \mathbf{S}}{\sum_{Y} \exp \boldsymbol{\theta}^T \mathbf{S}} \\
&= \mathbb{E}_{p_{\theta}(Y|Y^L,G)}\mathbf{S} - \mathbb{E}_{p_{\theta}(Y|G)}\mathbf{S}.
\end{aligned}
\tag{2.36}
$$

Another challenge here is that the graphical structure in PLP-FGM can be arbitrary and may contain cycles, which makes it intractable to directly calculate the expectation $\mathbb{E}_{p_{\theta}(Y|G)}\mathbf{S}$. A number of approximate algorithms have been proposed, such as Loopy Belief Propagation (LBP)

[111]. In this book, we utilize Loopy Belief Propagation. Specifically, we approximate marginal probabilities $p(y_i|\boldsymbol{\theta})$ and $p(y_i, y_j|\boldsymbol{\theta})$ using LBP. With the marginal probabilities, the gradient can be obtained by summing over all relationship nodes. It is worth noting that we need to perform the LBP process twice in each iteration, one time for estimating the marginal probability $p(y|G)$ and the other for $p(y|Y^L, G)$. Finally with the gradient, we update each parameter with a learning rate η. The learning algorithm is summarized in Algorithm 1.

Inferring Unknown Social Ties We now turn to describing how to infer the type of unknown social relationships. Based on learned parameters $\boldsymbol{\theta}$, we can predict the label of each relationship by finding a label configuration which maximizes the joint probability (Eq. (2.30)), i.e.,

$$Y^* = \mathrm{argmax}_{Y|Y^L} p(Y|G). \tag{2.37}$$

Again, we utilize the Loopy Belief Propagation (LBP) to compute the marginal probability of each relationship node $p(y_i|Y^L, G)$ and then predict the type of a relationship as the label with the largest marginal probability. The marginal probability is taken as the prediction confidence.

Time Complexity Analysis We use ν_1, ν_2, ν_3 to denote the number of attribute factors, correlation factors, and constraint factors in our PLP-FGM, respectively. In each round of LBP, the time cost of propagation is $\mathcal{O}(\nu_1 \cdot \dim(\boldsymbol{\Phi}) + \nu_2 \cdot \dim(\mathbf{g}) + \nu_3 \cdot \dim(\mathbf{h}))$, where $\dim(\cdot)$ is the dimension of a vector. We execute the learning algorithm for n iterations, and in each round we execute LBP for n_{LBP} iterations. Thus, we can estimate the time complexity as $\mathcal{O}((\nu_1 \cdot \dim(\boldsymbol{\Phi}) + \nu_2 \cdot \dim(\mathbf{g}) + \nu_3 \cdot \dim(\mathbf{h})) \times n \times n_{LBP})$.

Distributed Learning

As real social networks may contain millions of users and relationships, it is important for the learning algorithm to scale well with large networks. To address this issue, we develop a distributed learning method based on MPI (Message Passing Interface). The learning algorithm can be viewed as two steps: (1) compute the gradient for each parameter via loopy belief propagation; and (2) optimize all parameters with the gradient descents. The most expensive part is the step of calculating the gradient. Therefore, we develop a distributed algorithm to speed up the process.

We adopt a *master-slave* architecture, i.e., one master node is responsible for optimizing parameters, and the other slave nodes are responsible for calculating gradients. At the beginning of the algorithm, the graphical model of PLP-FGM is partitioned into P roughly equal parts, where P is the number of slave processors. This process is accomplished by graph segmentation software METIS [73]. The subgraphs are then distributed over slave nodes. Note that in our implementation, the edges (factors) between different subgraphs are eliminated, which results in an approximate solution. In each iteration, the master node sends the newest parameters $\boldsymbol{\theta}$ to all slaves. Slave nodes then start to perform Loopy Belief Propagation on the corresponding subgraph to calculate the marginal probabilities, then further compute the parameter gradient and send

Table 2.4: Data transferred in distributed learning algorithm

Phase	From	To	Data Description
Initialization	Master	Slave i	i-th subgraph
Iteration Beginning	Master	Slave i	Current parameters θ
Iteration Ending	Slave i	Master	Gradient in i-th subgraph

Table 2.5: Statistics of three data sets

Data set	Users	Unlabeled Relationships	Labeled Relationships
Publication	1,036,990	1,984,164	6,096
Email	151	3,424	148
Mobile	107	5,122	314

it back to the master. Finally, the master node collects and sums up all gradients obtained from different subgraphs, and updates parameters by the gradient descent method. The data transferred between the master and slave nodes are summarized in Table 2.4.

Evaluation

The proposed relationship mining approach is general and can be applied to many different scenarios. In this section, we present experiments on three different genres of data sets to evaluate the effectiveness and efficiency of our proposed approach. All data sets and codes are publicly available.[5]

Data Sets and Factor Definitions. We evaluate the proposed methods on three different data sets: Publication, Email, and Mobile. Statistics of the data sets are listed in Table 2.5.

- Publication. In the publication data set, we try to infer the advisor-advisee relationship from the co-author network. The data set is provided by [155]. Specifically, we have collected 1,632,442 publications from ArnetMiner [148] (from 1936–2010) with 1,036,990 authors involved. The ground truth is obtained in three ways: (1) manually crawled from researcher's homepage; (2) extracted from Mathematics Genealogy project;[6] and (3) extracted from AI Genealogy project.[7] In total, we have collected 2,164 advisor-advisee pairs as positive cases, and another 3,932 pairs of colleagues as negative cases. The mining results for advisor-advisee relationships are also available in the online system Arnetminer.org.

---

[5]http://arnetminer.org/socialtie/
[6]http://www.genealogy.math.ndsu.nodak.edu
[7]http://aigp.eecs.umich.edu

- Email. In the email data set, we aim to infer the manager-subordinate relationship from the email communication network. The data set consists of 136,329 emails between 151 Enron employees. The ground truth of manager-subordinate relationships is provided by [38].

- Mobile. In the mobile data set, we try to infer the friendship in mobile calling network. The data set is from Eagle et al. in [44]. It consists of call logs, bluetooth scanning logs, and location logs collected by a software installed in mobile phones of 107 users during a 10-month period. In the data set, users provide labels for their friendships. In total, 314 pairs of users are labeled as friends.

In the Publication data set, relationships are established between authors v_i and v_j if they co-authored at least one paper. For each pair of co-authors (v_i, v_j), our goal is to identify whether v_i is the advisor of author v_j. In this data set, we consider two types of correlations: (1) *Co-advisee*. The assumption is based on the fact that one could have only a limited number of advisors in her/his research career. Based on this, we define a correlation factor h_1 between nodes r_{ij} and r_{kj}. (2) *Co-advisor*. Another observation is that if v_i is the advisor of v_j (i.e., $r_{ij} = 1$), then v_i is very likely to be the advisor of some other student v_k who is similar to v_j. We define another factor function h_2 between nodes r_{ij} and r_{ik}.

In the Email data set, we try to discover the "manager-subordinate" relationship. A relationship (v_i, v_j) is established when two employees have at least one email communication. In total, there are 3,572 relationships among which 148 are labeled as manager-subordinate relationships. We try to identify the relationship types from the email traffic network. For example, if most of an employee's emails were sent to the same person, then the recipient is very likely to be her manager. A correlation named *co-recipient* is defined, that is, if a user v_i sent more than ϑ emails of which recipients including both v_j and v_k (ϑ is a threshold and is set as 10 in our experiment), then the relationship r_{ij} and r_{ik} are very likely to be the same. Therefore, a correlation factor is added between the two relationships. Two constraints named *co-manager* and *co-subordinate* are also introduced in an analogous way as that for the publication data.

In the Mobile data set, we try to identify whether two users have a friendship if there were at least one voice call or one text message sent from one to the other. Two kinds of correlations are considered. (1) *Co-location*. If more than three users arrived at the same location roughly the same time, we establish correlations between all the relationships in this groups. (2) *Related-call*. When v_i makes a call to both v_k and v_j from the same location, or makes a call to v_k immediately after the call with v_j, we add a related-call correlation factor between r_{ij} and r_{ik}.

In addition, we also consider some other factors in the three data sets. A detailed description of the factor definition for each data set is given in Table 2.6. Specifically, in the Publication data set, we define five categories of attribute factors: Paper count, Paper ratio, Co-author ratio, Conference coverage, and First-paper-year-diff. The definitions of the attributes are summarized in Table 2.6. In the Email data set, traffic-based features are extracted. For a relationship, we compute the number of emails for different communication types. In the Mobile data set, the

Table 2.6: Attributes used in the experiments. In the Publication data set, we use P_i and P_j to denote the set of papers published by author v_i and v_j, respectively. For a given relationship (v_i, v_j), five categories of attributes are extracted. In the Email data set, for relationship (v_i, v_j), number of emails for different communication types are computed. In the Mobile data set, the attributes are from the voice call/message/proximity logs

Data set	Factor	Description	
Publication	Paper count	$|P_i|, |P_j|$	
	Paper ratio	$|P_i|/|P_j|$	
	Coauthor ratio	$|P_i \cap P_j|/|P_i|, |P_i \cap P_j|/|P_j|$	
	Conference coverage	The proportion of the conferences which both v_i and v_j attended among conferences v_j attended.	
	First-paper-year-diff	The difference in year of the earliest publication of v_i and v_j.	
Email	Traffics	Sender	Recipients Include
		v_i	v_j
		v_j	v_i
		v_i	v_k and not v_j
		v_j	v_k and not v_i
		v_k	v_i and not v_j
		v_k	v_j and not v_i
		v_k	v_i and v_j
Mobile	#voice calls	The total number of voice call logs between two users.	
	#messages	Number of messages between two users.	
	Night-call ratio	The proportion of calls at night (8pm to 8am).	
	Call duration	The total duration time of calls between two users.	
	#proximity	The total number of proximity logs between two users.	
	In-role proximity ratio	The proportion of proximity logs in "working place" and in working hours (8am to 8pm).	

attributes we extracted are #voice calls, #messages, Night-call ratio, Call duration, #proximity, and In-role proximity ratio.

Evaluation Measures. To quantitatively evaluate the proposed method, we consider two aspects: performance and scalability. For the relationship mining performance, we consider two-fold cross-validation (i.e., half training and half testing) and evaluate the approaches in terms of accuracy, precision, recall, and F1-score. For scalability, we examine the execution time of the model learning. We compare our approach with the following methods for inferring relationship types:

SVM: It uses the relationship attribute vector x_i to train a classification model and predict the relationships by employing the classification model. We use the SVM-light package to implement SVM.

TPFG: It is an unsupervised method proposed in Wang et al. [155] for mining advisor-advisee relationships in publication network. This method is domain-specific and thus we only compare it with the Publication data set.

PLP-FGM-S: The proposed PLP-FGM is based on the partially labeled network. An alternative strategy is to train the model (parameters) with the labeled nodes only. We use this method to evaluate the necessity of the partial learning.

All the codes are implemented in C++, and all experiments are conducted on a server running Windows Server 2008 with Intel Xeon CPU E7520 1.87 GHz (16 cores) and 128 GB memory. The distributed learning algorithm is implemented on MPI (Message Passing Interface).

Table 2.7: Performance of relationship mining with different methods on three data sets: Publication, Email, and Mobile (%)

Data set	Method	Accuracy	Precision	Recall	F1-score
Publication	SVM	76.6	72.5	54.9	62.1
	TPFG	81.2	82.8	**89.4**	86.0
	PLP-FGM-S	84.1	77.1	78.4	77.7
	PLP-FGM	**92.7**	**91.4**	87.7	**89.5**
Email	SVM	82.6	79.1	**88.6**	83.6
	PLP-FGM-S	85.6	85.8	85.6	85.7
	PLP-FGM	**88.0**	**88.6**	87.2	**87.9**
Mobile	SVM	80.0	**92.7**	64.9	76.4
	PLP-FGM-S	80.9	88.1	71.3	78.8
	PLP-FGM	**83.1**	89.4	**75.2**	**81.6**

Accuracy Performance. Table 2.7 lists the accuracy performance of inferring the type of social relationships by the different methods.

Performance Comparison. Our method consistently outperforms other comparative methods on all the three data sets. In the Publication data set, PLP-FGM achieves a +27% (in terms of F1-score) improvement compared with SVM, and outperforms TPFG by 3.5% (F1-score) and 11.5% in terms of accuracy. We observe that TPFG achieves the best recall among all the four methods. This is because TPFG tends to predict more positive cases (i.e., inferring more advisor-advisee relationships in the co-author network), thus hurting the precision. As a result, TPFG underperforms our method by 8.6% in terms of precision. In Email and Mobile data set, PLP-FGM outperforms SVM by +4% and +5%, respectively.

Unlabeled Data Offers Improvement. From the result, it is clearly shown that by utilizing the unlabeled data, our model indeed obtains a significant improvement. Without using the unlabeled data, our model (PLP-FGM-S) results in a large performance reduction (−11.8% in terms of F1-score) on the publication data set. On the other two data sets, we also observe a clear performance reduction.

Table 2.8: Factor contribution analysis on three data sets (%)

Data set	Factors used	Accuracy	Precision	Recall	F1-score
Publication	Attributes	77.1	71.1	59.8	64.9
	+ Co-advisor	83.5	80.9	69.8	75.0 (+10.1%)
	+ Co-advisee	83.1	79.7	70.2	74.7 (+9.8%)
	All	92.7	91.4	87.7	89.5(+24.6%)
Email	Attributes	80.1	79.5	81.2	80.3
	+ Co-recipient	80.8	81.5	79.7	80.6 (+0.3%)
	+ Co-manager	83.1	82.8	83.5	83.2 (+2.9%)
	+ Co-subordinate	85.0	84.4	85.7	85.0 (+4.7%)
	All	88.0	88.6	87.2	87.9 (+7.6%)
Mobile	Attributes	81.8	88.6	73.3	80.2
	+ Co-location	82.2	89.2	73.3	80.4 (+0.2%)
	+ Related-call	81.8	88.6	73.3	80.2 (+0.0%)
	All	83.1	89.4	75.2	81.6 (+1.4%)

Factor Contribution Analysis. We perform an analysis to evaluate the contribution of different factors defined in our model. We first remove all the correlation/constraint factors and only keep the attribute factor, and then add each of the factors into the model and evaluate the performance improvement by each factor. Table 2.8 shows the result of factor analysis. We see that almost all the factors are useful for inferring the social relationships, but the contribution is very different. For example, for inferring the manager-subordinate relationship, the co-subordinate factor is the most useful factor which achieves a 4.7% improvement by F1-score, and the co-manager factor achieves a 2.9% improvement, while the co-recipient factor only results in a 0.3% improvement. By combining all the factors together, we can further obtain a 2.9% improvement.

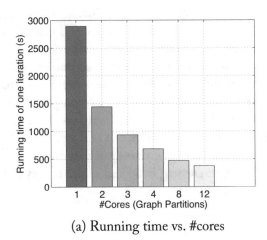

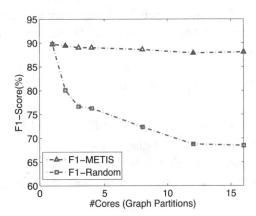

(a) Running time vs. #cores

(b) Speedup vs. #cores

Figure 2.4: Scalability performance.

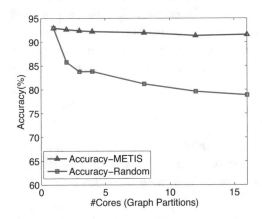

Figure 2.5: Approximation of graph partition.

An extreme phenomenon appears on the Mobile data set. With each of the two factors (co-location and related-call), we cannot obtain a clear improvement (0.2% and 0.0% by F1). However, when combining the two factors and the attribute factor together, we can achieve a 1.4% improvement, 7 times higher than that obtained by the separated case. This is because our model not only considers different factors, but also considers the correlation between them.

Scalability Performance. We conduct a series of experiments to evaluate the scalability performance of our distributed learning algorithm on the Publication data set. Figure 2.4 shows the running time and speedup of the distributed algorithm with different number of computer nodes

(2,3,4,8,12 cores) used. The speedup curve is close to the perfect line at the beginning. Although the speedup inevitably decreases when the number of cores increases, it can achieve $\sim 8\times$ speedup with 12 cores. It is noticeable that the speedup curve is beyond the perfect line when using 4 cores, it is not strange since our distributed strategy is approximated. In our distributed implementation, graphs are partitioned into subgraphs, and the factors across different parts are discarded. Thus, the graph processed in distributed version contains less edges, making the computational cost less than the amount in the original algorithm. The effect of subgraph partition is illustrated in Figure 2.5. By using good graph partition algorithm such as METIS, the performance only decreases slightly (1.4% in accuracy and 1.6% in F1-score). A theoretical study of the approximate ratio for the distributed learning algorithm would be an interesting issue and is also one of our ongoing work.

Summary
In this section, we study the problem of inferring the type of social ties in large networks. We propose a partially labeled pairwise factor graph model (PLP-FGM) to learn to infer the relationship semantics. In PLP-FGM, relationships in social network are modeled as nodes, the attributes, correlations, and global constraints are modeled as factors. An efficient algorithm is proposed to learn model parameters and predict unknown relationships. Experimental results on three different types of data sets validate the effectiveness of the proposed model. To further scale up to large networks, a distributed learning algorithm is developed. Experiments demonstrate good parallel efficiency of the distributed learning algorithm.

2.3.4   ACTIVELY LEARNING TO INFER SOCIAL TIES

Another important question is how we can learn the mapping function f effectively. In many situations, labeled data is limited and expensive. The problem is, can we design a strategy to *actively* learn the model with minimal labeling cost? Formally, we have the following.

Problem 2.7   Active social tie inference. Given a partially labeled network $G = (V, E^L, E^U, R^L, \mathbf{W})$, and a labeling budget b (number of user interactions), our objective is to select a subset of unknown relationships $A \subset E^U$ within the constraint of b to label, so that the performance of predictive function f can be maximally improved.

Our formulation of inferring social relationships is very different from existing works on relation mining [23], which focuses on detecting the relationships from the content information, while we focus on mining relationship semantics in social networks. Diehl et al. [38] and Wang et al. [155] investigated the problem of relationship identification. However, they studied the problem in specific domains (Email network or Publication network). Backstrom et al. [6] proposed an algorithm based on supervised random walks for link prediction. Crandall et al. [32] incorporated geographic coincidences to infer social ties. Different from these works which aim at link prediction, our goal is to infer the types of relationships. There are also works on inferring the types of relationships. Hopcroft et al. [70] explored the problem of reciprocal relationship prediction and Tang et al. [142] developed a framework for classifying the type of social relationships by learning across heterogeneous networks. Yang et al. [168] studied the retweeting behavior. Leskovec et al. [92] focused on the prediction of edge signs (positive or negative). However, they do not consider how to make optimal use of user interaction.

Formally, for actively selecting helpful relationships to query the user, we define a quality function $Q(A)$, which measures the expected improvement of the prediction performance by labeling relationships in set A. The problem can be then defined as an optimization problem of $Q(A)$, i.e.,

$$A^* = \arg \max_{A \subset Y^U} Q(A), |A| = b, b > 0.$$

To quantify $Q(A)$, we could consider how a selected node can influence the others. For example, correction of a centered relationship may trigger a spread of the correction, thus helping infer correlated relationships.

Based on the above intuitions, we develop an Influence-Maximization Selection (IMS) model and a Belief-Maximization Selection (BMS) model for actively inferring the types of social relationships. The IMS model selects the most influential nodes, by leveraging the network structure and the uncertainty obtained from PLP-FGM. The BMS model further incorporates the active selection process into the learning process of PLP-FGM.

Baseline Methods

The quality function $Q(A)$ can be defined in different forms. Without any constraints, optimizing the quality function $Q(A)$ needs to enumerate all possible subsets $A \subset Y^U$, which is obviously NP-hard. We first review two baseline greedy algorithms.

Maximum Uncertainty (MU). A most common selection strategy for active learning is to select the most uncertain relationships. The uncertainty of an unlabeled relationship y_i is measured by the *entropy* $H(y_i) = -\sum_{y \in \mathcal{y}} p(y_i = y) \log p(y_i = y)$. Based on this intuition, we can define the quality function as

$$Q_{MU}(A) = H(A) \tag{2.38}$$

where $H(A) = \sum_{y_i \in A} H(y_i)$.

Information Density (ID). A drawback of the Maximum Uncertainty strategy is its tendency to choose outliers. Thus, we employ another strategy, Information Density, proposed in Settles and Craven [129]. The idea is to choose the most *representative* nodes in Y^U, which are supposed to be the most informative ones. Based on this intuition, we measure the informativeness of a node by its cosine similarity to all other unlabeled nodes in the sense of the attributes attached to a node. Formally, we define the quality function as

$$Q_{ID}(A) = \sum_{i \in A} H(y_i) \times \left[\frac{1}{|Y^U|} \sum_{j \in Y^U} sim(\mathbf{x}_i, \mathbf{x}_j) \right] \tag{2.39}$$

where $sim(\mathbf{x}_i, \mathbf{x}_j) = \frac{\mathbf{x}_i \cdot \mathbf{x}_j}{\|\mathbf{x}_i\| \times \|\mathbf{x}_j\|}$. Note that we again employ the entropy of a relationship node $H(y_i)$ to leverage the "base" informativeness.

Proposed Methods for Actively Learning PLP-FGM

We propose two new algorithms, i.e., Influence-Maximization Selection model (IMS) and Belief-Maximization Selection model (BMS), for actively learning the presented PLP-FGM model.

Influence-Maximization Selection (IMS). All the strategies mentioned above do not consider the network structure information. As relationship nodes in PLP-FGM are correlated, the most influential nodes are more likely to help improve the overall performance of the model. Existing work has studied several influence propagation models, including the Linear Threshold Model (LTM) in Kempe et al. [79]. The LTM model sets a threshold value ε_i for each node, and weights $b_{i,j}$ for its edges, satisfying $\sum_{j \in NB(i)} b_{i,j} \leq 1$. In each time stamp, if $\sum_{j \in NB(i) \wedge activated(j)} b_{i,j} \geq \varepsilon_i$, then the node i will be activated. We develop a variation of the LTM by incorporating a score for each node reflecting the strength of the influence spreading in our model. The propagation process is described as follows.

- The graph is the same as the PLP-FGM model. In addition, we call a relationship node as "activated" when its label y_i is determined. The initial activated set of nodes is Y^L. We assign a threshold $\varepsilon_i = \sum_{y \in \mathcal{Y}} |p(y_i = y|G, Y^L) - \frac{1}{|\mathcal{Y}|}|$ for each node. Thus, a node with higher uncertainty will be easier to be activated.

- When a node i is activated, it spreads its gained score increment $(g_i - \varepsilon_i)$ to its neighbor nodes $j \in NB(i)$ with a weight $b_{i,j}$, i.e., $g_j \leftarrow g_j + b_{i,j}(g_i - \varepsilon_i)$. The gained score increment reflects the improvement of confidence brought by user labeling, therefore the influence by labeling an uncertain relationship will be greater than labeling a more certain relationship. To simplify the problem, we set weight $b_{i,j} = 1/|NB(j)|$.

- If a node is labeled by the user, we set it as activated and assign its gained score as 1. The gained score for other nodes is set to 0 at the beginning. Once an inactivated node k gains a score which exceeds the threshold, i.e., $g_k > \varepsilon_i$, it will become activated and spread its gained score similarly. An activated node only spreads its gained score once and remains its status.

We define the quality function $Q_{IMS}(A)$ as the total number of activated nodes after the propagation process. Finding the set A that maximizes the quality function $Q_{IMS}(A)$ is NP-hard.

Similar to [79], in this chapter, we use a greedy strategy to approximate the solution. Note that unlike the LTM, we do not guarantee a lower bound of error for the greedy optimization method.

Belief-Maximization Selection (BMS). To quantify the influence of one node on the others, we employ the *belief* of each node obtained by Loopy Belief Propagation in our model. We define a heuristic by removing the effect of attributes from the belief score, denoted by $\mathcal{B}(y_i|G, Y^L)$. More precisely,

$$\mathcal{B}(y_i|G, Y^L) = \exp\{\boldsymbol{\theta}^T \mathbf{s}(y_i) - \boldsymbol{\lambda}^T \boldsymbol{\Phi}(y_i, \mathbf{x}_i)\}.$$

By normalizing the belief of one relationship node, we obtain the *belief marginal probability*.

$$p_{\mathcal{B}}(y_i|G, Y^L) = \frac{1}{Z_{\mathcal{B}}}\mathcal{B}(y_i|G, Y^L),$$

where $Z_{\mathcal{B}}$ is the normalization factor. It estimates the marginal probability distribution of a relationship node where the information of its attribute vector is absent.

A basic intuition is, the belief of a relationship node is *monotonically increasing* with respect to the number of relationship nodes of the same type, i.e., $\mathcal{B}(y_i = y|G, Y^L)$ is monotonically increasing with respect to the number of relationships with label y.[8] Without loss of generality, we first consider the binary relationship mining problem, i.e., there are only two possible labels of relationships ($\mathcal{Y} = \{0, 1\}$). In the binary setting, we further consider the active selection for each type separately. This is because when mixing the different types of relationships together, it cannot be guaranteed to have a closed-form solution. Thus, when users provide only positive feedback, our objective is to find a set of positive nodes. Accordingly, we define the quality function of the *positive-oriented BMS strategy* as:

$$Q_{BMS+}(A) = \sum_{y_i \in Y_{(1)}^U} p_{\mathcal{B}}(y_i = 1|G, Y^L \cup A), \tag{2.40}$$

where $Y_{(1)}^U = \{y_i|y_i \in Y^U \wedge \mathcal{B}(y_i = 1|G, Y^L) \geq \mathcal{B}(y_i = 0|G, Y^L)\}$.

Symmetrically, if the users provide only *negative* feedback, we can adopt a *negative-oriented BMS strategy*, with the following quality function:

$$Q_{BMS-}(A) = \sum_{y_i \in Y_{(0)}^U} p_{\mathcal{B}}(y_i = 0|G, Y^L \cup A). \tag{2.41}$$

---

[8]We present a sufficient condition for this assumption. If for all $y' \in \mathcal{Y}, y' \neq y$, we can have $\boldsymbol{\alpha}^T \exp\{\mathbf{g}(y, y) + \boldsymbol{\beta}^T \mathbf{h}(y, y)\} \geq \exp\{\boldsymbol{\alpha}^T \mathbf{g}(y, y') + \boldsymbol{\beta}^T \mathbf{h}(y, y')\}$, then $\mathcal{B}(y_i = y|G, Y^L)$ is monotonically increasing with respect to the number of y-labeled relationships in Y^L.

Input: G, b

Output: a set of selected relationships A

Train PLP-FGM and get the parameter configuration $\boldsymbol{\theta}$;

$A^+, A^- \leftarrow \emptyset$;

for $b/2$ *times* **do**

    Use Loopy Belief Propagation (LBP) to obtain the probability distribution for each relationship;

    Find $y_{\max+} = \mathrm{argmax}_{y_i \in Y^U} \, p(y_i = 1 | G, Y^L)(Q_{BMS+}(A^+ \cup y_i) - Q_{BMS+}(A^+))$;

    Move $y_{\max+}$ from Y^U to A^+;

    $Y^L \leftarrow Y^L \cup A^+$;

    Use Loopy Belief Propagation (LBP) to obtain the probability distribution for each relationship;

    Find $y_{\max-} = \mathrm{argmax}_{y_i \in Y^U} \, p(y_i = 0 | G, Y^L)(Q_{BMS-}(A^- \cup y_i) - Q_{BMS-}(A^-))$;

    Move $y_{\max-}$ from Y^U to A^-;

    $Y^L \leftarrow Y^L \cup A^-$;

end

Algorithm 2: Belief-maximization selection.

The optimization of both quality functions $Q_{BMS+}(A)$ and $Q_{BMS-}(A)$ is NP-hard. However, as both quality functions are submodular (theoretical analysis is given in Section 2.3.4), a solution with an approximation ratio of $(1 - 1/e)$ can be obtained using a greedy algorithm: at each time, it selects the relationship which is expected to provide the maximum marginal increase of the quality function. Notice that we treat the examining relationship node y_i as if it is positive-labeled when optimizing $Q_{BMS+}(A)$, or negative-labeled for $Q_{BMS-}(A)$, since the active learning algorithm is label-unaware in the selection stage. In order to leverage the risk that a selected relationship is not labeled as expected, we employ a weighting factor $p(y_i | G, Y^L)$ to reflect how likely the relationship would be labeled as positive (negative).

To prevent making an imbalance selection, we intuitively use Q_{BMS+} to choose $b/2$ nodes (where b is the number of relationships we expect to query the user each time), and then use Q_{BMS-} for the rest. Algorithm 2 formally describes the selection process. This selection strategy is denoted by BMS. Note that BMS combines both BMS+ and BMS-. Thus, it cannot guarantee a lower error bound of the approximation.

Theoretical Analysis

We give a theoretical analysis of proposed active learning models. The approximation ratio of the IMS model is given in Kempe et al. [79]. Here, we focus on the proof of approximation guarantees of the BMS model. The proof is based on the submodular property, which indicates that the marginal gain from adding an element to a set S is at least as high as the marginal gain from adding the same element to a superset of S. The following is a formal definition of the submodular set function.

Definition 2.8   (Submodular) A set function F defined on set S is called submodular, if for all $A \subset B \subset S$ and $s \notin B$, it satisfies $F(A \cup \{s\}) - F(A) \geq F(B \cup \{s\}) - F(B)$.

Given a submodular function F, which is also monotone and non-negative, it is an NP-hard problem to find a k-element subset S to optimize F. But a greedy algorithm can result in an approximation ratio of $(1 - 1/e)$. It constructs the subset by selecting elements one at a time, each time choosing an element that provides the largest marginal increase in the function value. Thus, we have the following.

Theorem 2.9   *For a non-negative, monotone submodular function F, let S be the k-element subset decided by the following algorithm: for k times, each time choose an element which gives the maximum marginal increase of F and move it to S. Let S^* denotes the optimal solution. Then we have $F(S) \geq (1 - \frac{1}{e})F(S^*)$.*

Before we prove the submodularity of the quality function Q_{BMS+}, we first prove the monotonicity of function $p_B(y_i = 1|S)$.

Lemma 2.10   *For all $y_i \in Y^U$, function $p_B(y_i = 1|S)$ is monotonic with respect to S.*

Suppose x is another unlabeled relationship. We have

$$
\begin{aligned}
p_B(y_i = 1|S \cup \{x\}) &= \frac{\mathcal{B}(y_i = 1|S \cup \{x\})}{\mathcal{B}(y_i = 1|S \cup \{x\}) + \mathcal{B}(y_i = 0|S \cup \{x\})} \\
&= \frac{1}{1 + \frac{\mathcal{B}(y_i = 0|S \cup \{x\})}{\mathcal{B}(y_i = 1|S \cup \{x\})}}.
\end{aligned}
$$

Let $k_1 = \frac{\mathcal{B}(y_i = 0|S \cup \{x\})}{\mathcal{B}(y_i = 1|S \cup \{x\})}$, then

$$
p_B(y_i = 1|S \cup \{x\}) = \frac{1}{1 + k_1}.
$$

Similarly, let $k_2 = \frac{\mathcal{B}(y_i = 0|S)}{\mathcal{B}(y_i = 1|S)}$, then

$$
p_B(y_i = 1|S) = \frac{1}{1 + k_2}.
$$

According to the assumption in 2.3.4, it is obvious that $k_1 \leq k_2$. Obviously,

$$p_B(y_i = 1|S \cup \{x\}) \geq p_B(y_i = 1|S).$$

Now we prove the submodularity of the quality function Q_{BMS+} defined by Equation (2.40).

Theorem 2.11 *The quality function $Q_{BMS+}(S)$ satisfies the submodular property, when $S \subset Y_{(1)}^U$.*

Proof. The first step is to prove that function $F(S) = p(y_i = 1|G, Y^L \cup S)$ is submodular with respect to S. Suppose $A \subset B \subset Y_{(1)}^U$, and there is another unlabeled relationship $x \notin B$. Similarly, we define k_1, k_2, k_3, k_4 below:

$$k_1 = \frac{B(y_i = 0|G, Y^L \cup A \cup \{x\})}{B(y_i = 1|G, Y^L \cup A \cup \{x\})}, \quad k_2 = \frac{B(y_i = 0|G, Y^L \cup A)}{B(y_i = 1|G, Y^L \cup A)}$$

$$k_3 = \frac{B(y_i = 0|G, Y^L \cup B \cup \{x\})}{B(y_i = 1|G, Y^L \cup B \cup \{x\})}, \quad k_4 = \frac{B(y_i = 0|G, Y^L \cup B)}{B(y_i = 1|G, Y^L \cup B)}.$$

Since $A \subset B \subset Y_{(1)}^U$, we have $k_1, k_2, k_3, k_4 \leq 1$. In addition,[9] since our factor functions are defined as exponential-linear functions, we can have $k_1/k_2 = k_3/k_4$. We define α and β as follows:

$$\alpha = \frac{k_1}{k_2} = \frac{k_3}{k_4} \leq 1, \beta = \frac{k_3}{k_1} = \frac{k_4}{k_2} \leq 1.$$

Then we can obtain the following inequality:

$$
\begin{aligned}
\delta(A, x) &= p_B(y_i = 1|G, Y^L \cup A \cup \{x\}) - p_B(y_i = 1|G, Y^L \cup A) \\
&= \frac{1}{1 + k_1} - \frac{1}{1 + k_2} \\
&= \frac{(1 - \alpha)k_2}{(1 + \alpha k_2)(1 + k_2)} \\
\delta(B, x) &= p_B(y_i = 1|G, Y^L \cup B \cup \{x\}) - p_B(y_i = 1|G, Y^L \cup B) \\
&= \frac{(1 - \alpha)k_4}{(1 + \alpha k_4)(1 + k_4)} \\
&= \delta(A, x)\frac{(1 + \alpha)\beta k_2 + \beta + \alpha\beta k_2^2}{(1 + \alpha)\beta k_2 + 1 + \alpha\beta^2 k_2^2} \\
&\leq \delta(A, x)
\end{aligned}
$$

[9]If there is no factor function between x and y_i, the conclusion is obvious; otherwise, $B(y_i|G, Y^L \cup S \cup \{x\})/B(y_i|G, Y^L \cup S)$ is only relevant to the factor function between x and y_i since relationships in S remain their labels, and the conclusion can be derived accordingly.

Then we give the proof of the submodularity of quality function Q_{BMS+}. Suppose $A \subset B \subset Y^U$, and there is another unlabeled relationship $y_x \notin B$.

$$
\begin{aligned}
\Delta(A, x) &= Q_{BMS+}(A \cup \{y_x\}) - Q_{BMS+}(A) \\
&= \sum_{y_i \in Y^U_{(1)}} p_B(y_i = 1 | G, Y^L \cup A \cup \{y_x\}) - \sum_{y_i \in Y^U_{(1)}} p_B(y_i = 1 | G, Y^L \cup A) \\
&= \sum_{y_i \in Y^U_{(1)} \backslash (A \cup \{y_x\})} [\delta(A, y_x)] + 1 - p_B(y_x = 1 | G, Y^L \cup A) \\
&\geq \sum_{y_i \in Y^U_{(1)} \backslash (B \cup \{y_x\})} [\delta(B, y_x)] + 1 - p_B(y_x = 1 | G, Y^L \cup B) \\
&= \Delta(B, x).
\end{aligned}
$$

Therefore, we have proved that Q_{BMS+} is submodular. The submodularity of Q_{BMS-} can be proved in a similar way. According to Theorem 2.9, it guarantees a lower bound of the greedy algorithm employed for the BMS model. □

Evaluation

We still use the data sets used in Section 2.3.3 to evaluate the different active learning algorithms. More specifically, in each data set, we first randomly select 10 relationships as the initial labeled set Y^L. And then we iteratively perform the active selection algorithm, each time selecting $b = 10$ relationships to query. After each round of selection, we learn the PLP-FGM model and evaluate the prediction performance. We implement the experiment for ten times on each data set and use the mean of F1-score for evaluation.

Comparison Methods. We consider the following baseline methods.[10]

Random: It randomly selects b nodes in Y^U at each time.

Maximum Uncertainty (MU): It chooses the most b uncertain nodes among unlabeled relationships Y^U.

Information Density (ID): It chooses b nodes with the maximum average similarity to all other nodes in Y^U, proposed in Settles and Craven [129].

Effect of Active Learning. We plot the learning curves on each data set in Figure 2.6, and list the average F1-score by all selection strategies in Table 2.9. The results clearly demonstrate the effectiveness of the active selection strategies. In the Publication data set, the overall F1-score of the IMS strategy with 100 samples labeled outperforms the Random algorithm by +7.4%. In Email

---

[10]We did not consider the *co-advisee* correlation in the model when dealing with the Publication data set and the *co-subordinate* correlation for the Email data set, since they conflict with the assumption of monotonic belief in Section 2.3.4.

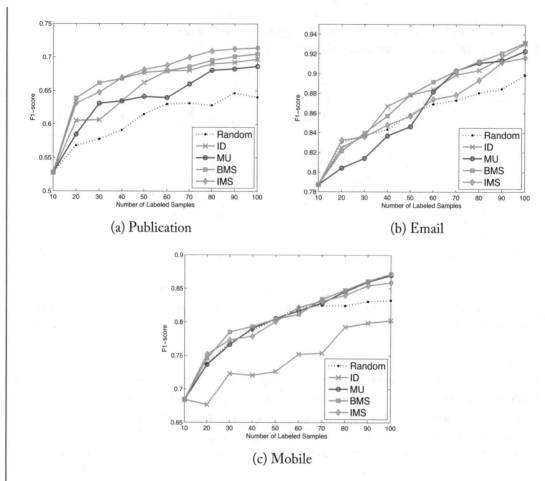

(a) Publication

(b) Email

(c) Mobile

Figure 2.6: Learning curves in terms of F1-score.

and Mobile data set, the BMS strategy achieves the best performance, with an improvement of +3.3% and +3.9%, respectively.

Performance Comparison. In the publication data set, both the proposed BMS and IMS strategy significantly perform better than all the baseline methods (paired t-tests with 95% significance). BMS also significantly outperforms Random and ID strategy in Mobile data set, while its performance is close to MU. The performance of IMS is shown better than ID in Mobile data set, but is close to other baseline methods. In Email data set, BMS significantly outperforms Random, while the performance of other methods seems close to each other. Generally, the proposed BMS strategy performs more consistently, and obtains better results in two of the three data sets.

Table 2.9: Average F1-score by all selection strategies (%). The results were obtained by randomly selecting 10 relationships as the initial labeled set Y^L, and then iteratively perform the active selection algorithms, each time with $b = 10$ relationships to query

Data set	Random	MU	ID	BMS	IMS
Publication	60.6	63.7	64.8	66.4	**66.8**
Email	85.6	86.2	87.3	**87.6**	86.3
Mobile	79.2	80.0	74.3	**80.4**	79.9

The performance of IMS strategy is the best in Publication data set, but seems close to baseline methods in the other two data sets.

Network Information Helps. According to factor contribution analysis mentioned before, co-advisor factor in Publication data set contributes the most. This explains why the proposed methods achieve better performance than the alternative baseline methods in Publication data set. The average F1-score of BMS and IMS reaches 65% with less than 30 labeled samples, while ID uses more than 40, and MU uses more than 60. In Email and Mobile data set, BMS still takes advantage of the network information, but the improvement shrinks due to the considerable decrease of factor contribution.

In-Depth Analysis of BMS. There are also some variations of BMS and we conduct a comparison between them. BMS+ selects all b nodes optimizing $Q_{BMS+}(A)$, while BMS- employs $Q_{BMS-}(A)$. Figure 2.7 shows the average F1-score of the different versions. In Publication and Email data set, the difference between BMS and BMS+ is minor, while the performance of BMS- drops. It might be resulted from different criteria of these three strategies. BMS+ tends to obtain true-positive samples, whereas BMS- is more likely to acquire true-negative samples. F1-score excludes the impact of true-negative samples, and therefore undermines the performance of BMS-. The gap disappears in Mobile data set, probably due to the weak contribution of its correlation and constraint factors.

Summary

In this section, we explain how to utilize active learning to help infer social ties. We propose two active learning strategies: Influence-Maximization Selection and Belief-Maximization Selection, both aiming to capture the inter-relationship influence. Experimental results show that BMS and IMS often achieve significant better performance than baseline methods.

2.3.5   INFERRING SOCIAL TIES ACROSS HETEROGENEOUS NETWORKS

In the previous sections, we introduced methodologies to infer particular types of relationships in different specific social networks. Now we discuss how to generalize the problem to infer social ties across multiple heterogeneous networks.

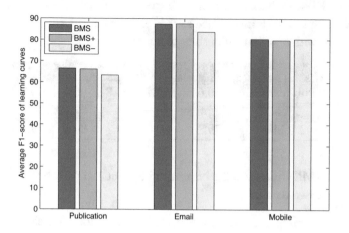

Figure 2.7: Performance comparison between variations of BMS.

In traditional methods, sufficient labeled relationships are usually necessary to learn a good predictive model for inferring social ties. However, the availabilities of labeled relationships in different networks are very unbalanced. In some networks, such as Slashdot, it might be easy to collect the labeled relationships (e.g., trust/distrust relationships between users), while in most other networks, it may be difficult (or even infeasible) to obtain the labeled information. A challenging question is: can we leverage the labeled relationships from one network to infer the type of relationships in another totally different network?

Problem Formulation. Figure 2.8 gives an example of inferring social ties across a product-reviewer network and a mobile communication network. In Figure 2.8, the left sub-figure is the input to our problem: a reviewer network, which consists of reviewers and relationships between reviewers, and a mobile network, which consists of mobile users and their communication relationships (via calling or texting message). The right sub-figure shows the output of our problem: the inferred social ties in the two networks. In the reviewer network, we infer the trust/distrust relationships and in the communication network, we identify friendships, colleagues, and families. The middle of Figure 2.8 is the component of knowledge transfer for inferring social ties in different networks. This is the key objective of this work. The fundamental challenge is how to bridge the available knowledge from different networks to help infer the different types of social relationships.

Formally, let $G = (V, E^L, E^U, \mathbf{X})$ denote a partially labeled social network, where E^L is a set of labeled relationships and E^U is a set of unlabeled relationships with $E^L \cup E^U = E$; $\mathbf{X}$ is an $|E| \times d$ attribute matrix associated with edges in E with each row corresponding to an edge, each column an attribute, and an element x_{ij} denoting the value of the jth attribute of edge e_i.

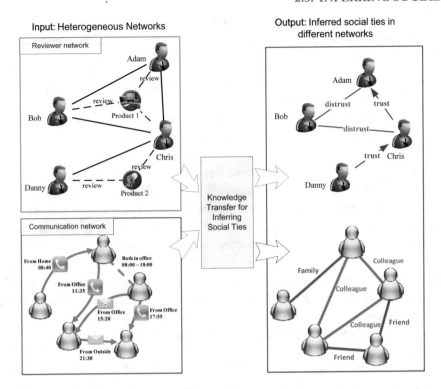

Figure 2.8: Example of inferring social ties across two heterogeneous networks: a product-reviewer network and a mobile communication network.

The label of edge e_i is denoted as $y_i \in \mathcal{Y}$, where $\mathcal{Y}$ is the possible space of the labels (e.g., family, colleague, classmate).

Given this, the input to our problem consists of two partially labeled networks G_S (source network) and G_T (target network) with $|E_S^L| \gg |E_T^L|$ (with an extreme case of $|E_T^L| = 0$). Please note that the two networks might be totally different (with different sets of vertexes, i.e., $V_S \cap V_T = \emptyset$, and different attributes defined on edges).

In real social networks, the relationship could be undirected (e.g., friendships in a mobile network) or directed (e.g., manager-subordinate relationships in an enterprise email network). To keep things consistent, we will concentrate on the undirected network. In addition, the label of a relationship may be static (e.g., the family-member relationship) or change over time (e.g., the manager-subordinate relationship). In this work, we focus on static relationships.

Given a source network G_S with abundantly labeled relationships and a target network G_T with a limited number of labeled relationships, the goal is to learn a predictive function $f : (G_T|G_S) \rightarrow Y_T$ for inferring the type of relationships in the target network by leveraging the supervised information (labeled relationships) from the source network.

Without loss of generality, we assume that for each possible type y_i of relationship e_i, the predictive function will output a probability $p(y_i|e_i)$; thus our task can be viewed as obtaining a triple $(e_i, y_i, p(y_i|e_i))$ to characterize each link e_i in the social network.

Data Sets. To study this problem, we try to find a number of different types of networks to investigate the problem of inferring social ties across heterogeneous networks. In this study, we consider five different types of networks: Epinions, Slashdot, Mobile, Coauthor, and Enron. Table 2.5 lists statistics of the five networks. All data sets and codes used in this work are publicly available.[11]

Epinions is a network of product reviewers. Each user on the site can post a review on any product and other users would rate the review with trust or distrust. In this data, we created a network of reviewers connected with trust and distrust relationships. The data set consists of 131,828 nodes (users) and 841,372 edges, of which about 85.0% are trust links. 80,668 users received at least one trust or distrust edge. Our goal on this data set is to infer the trust relationships between users.

Slashdot is a network of friends. Slashdot is a site for sharing technology related news. In 2002, Slashdot introduced the Slashdot Zoo which allows users to tag each other as "friends" (like) or "foes" (dislike). The data set is comprised of 77,357 users and 516,575 edges of which 76.7% are "friend" relationships. Our goal on this data set is to infer the "friend" relationships between users.

Mobile is a network of mobile users. The data set is from Eagle et al. [44]. It consists of the logs of calls, blue-tooth scanning data and cell tower IDs of 107 users during about 10 months. If two users communicated (by making a call and sending a text message) with each other or co-occurred in the same place, we create an edge between them. In total, the data contains 5,436 edges. Our goal is to infer whether two users have a friend relationship. For evaluation, all users are required to complete an online survey, in which 157 pairs of users are labeled as friends.

Coauthor is a network of authors. The data set, crawled from ArnetMiner.org [148], is comprised of 815,946 authors and 2,792,833 co-author relationships. In this data set, we attempt to infer advisor-advisee relationships between co-authors. For evaluation, we created a smaller ground truth data in the following ways: (1) collecting the advisor-advisee information from the Mathematics Genealogy project;[12] and the AI Genealogy project[13] (2) manually crawling the advisor-advisee information from researchers' homepages. Finally, we have created a data set with 1,534 co-author relationships, of which 514 are advisor-advisee relationships. The data set was used in Wang et al. [155].

Enron is an email communication network. It consists of 136,329 emails between 151 Enron employees. Two types of relationships, i.e., manager-subordinate and colleague, were annotated between these employees. The data set was provided by Diehl et al. [38]. Our goal on

---

[11]http://arnetminer.org/socialtieacross/
[12]http://www.genealogy.math.ndsu.nodak.edu
[13]http://aigp.eecs.umich.edu

this data set is to infer manager-subordinate relationships between users. There are in total 3,572 edges, of which 133 are manager-subordinate relationships.

Please note that for the first three data sets (i.e., Epinions, Slashdot, and Mobile), our goal is to infer undirected relationships (friendships or trustful relationships), while for the other two data sets (i.e., Coauthor and Enron), our goal is to infer directed relationships (the source end has a higher social status than the target end, e.g., advisor-advisee relationships and manager-subordinate relationships).

Table 2.10: Statistics of five data sets

Relationship	Dataset	#Nodes	#Edges
Trust	Epinions	131,828	841,372
Friendship	Slashdot	77,357	516,575
Friendship	Mobile	107	5,436
Advisor-advisee	Coauthor	815,946	2,792,833
Manager-subordinate	Enron	151	3,572

Observations

As a first step, we engage in some high-level investigation of how different factors influence the formation of different social ties in different networks. Generally, if we consider inferring particular social ties in a specific network (e.g., mining advisor-advisee relationships from the Coauthor network), we can define domain-specific features and learn a predictive model based on labeled training data. The problem becomes very different, when handling multiple heterogeneous networks, as the defined features in different networks may be significantly different. To solve this problem, we connect our problem to several basic social psychological theories and focus our analysis on the network based correlations via the following statistics.

1. *Social balance [45].* How is the social balance property satisfied and correlated in different networks?

2. *Structural hole [21].* Would structural holes have a similar behavior pattern in different networks?

3. *Social status [34, 60, 93].* How do different networks satisfy the properties of social status?

4. *"Two-step flow" [89].* How do different networks follow the "two-step flow" of information propagation?

Social Balance. Social balance theory suggests that people in a social network tend to form into a balanced network structure. Figure 2.9 shows the probabilities of balanced triads of the three

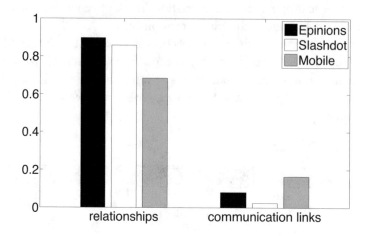

Figure 2.9: Social balance. Probabilities of balanced triads in different networks based on communication links and friendships (or trustful relationships). Based on communication links, different networks have very different balance probabilities (e.g., the balance probability in the mobile network is nearly seven times higher than that of the slashdot network). While based on friendships the three networks have relatively similar probabilities.

undirected networks (Epinions, Slashdot, and Mobile). In each network, we compare the probability of balanced triads based on communication links and that based on friendships (or trust relationships). For example, in the Mobile network, the communication links include making a call or sending a message between users. We find it interesting that different networks have very different balance probabilities based on the communication links, e.g., the balance probability in the mobile network is nearly seven times higher than that of the slashdot network, while based on friendships (or trustful relationships) the three networks have relatively similar balance probabilities (with a maximum of +28% difference).

Structural Hole. A person is said to span a *structural hole* in a social network if he or she is linked to people in parts of the network that are otherwise not well connected to one another [21]. Arguments based on structural holes suggest that there is an informational advantage to having friends in a network who do not know each other. Our idea here is to test if a structural hole tends to have the same type of relationship with the other users. We first employ a simple algorithm to identify structural hole users in a network. Following the informal description of structural holes [21], for each node, we count the number of pairs of neighbors who are not directly connected. All users are ranked based on the number of pairs and the top 1% users[14] with the highest numbers are viewed as structural holes in the network. Figure 2.10 shows the probabilities that two users (A and B) have the same type of relationship with another user (say C), conditioned on whether

[14]This is based on the observation that less than 1% of the Twitter users produce 50% of its content [165].

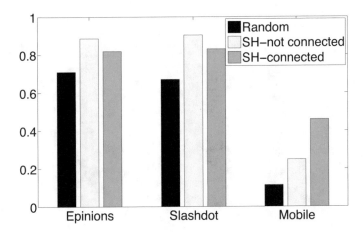

Figure 2.10: Structural hole. Probabilities that two connected (or disconnected) users (A and B) have the same type of relationship with user C, conditioned on whether user C spans a structural hole or not. It is clear that (1) users are more likely (averagely +70% higher than chance) to have the same type of relationship with C if C spans a structural hole; and (2) disconnected users are more likely than connected users to have the same type of relationship with a user who spans a structural hole (except the mobile network).

user C spans a structural hole or not. We have two interesting observations: (1) users are more likely (on average +70% higher than chance) to have the same type of relationship with C if C spans a structural hole; and (2) disconnected users are more likely than connected users to have the same type of relationship with a user classified as spanning a structural hole. One exception is the mobile network, where most mobile users in the data set are university students and thus friends frequently communicate with each other.

Social Status. Social status theory [34, 60, 93] is based on the directed relationship network. We conducted an analysis on the Coauthor and Enron networks, where we aim to find directed relationships (advisor-advisee and manager-subordinate). We found nearly 99% of triads in the two networks satisfy the social status theory, which was also validated in Leskovec et al. [93]. We investigate more by looking at the distribution of different forms of triads in the two networks. Specifically, there are in total 16 different forms of triads [93]. We select five most frequent forms of triads in the two networks. For easy understanding, given a triad (A, B, C), we use 1 to denote the advisor-advisee relationship and 0 colleague relationship, and three consecutive numbers 011 to denote A and B are colleagues, B is C's advisor and A is C's advisor. It is striking that although the two networks (Coauthor and Enron) are totally different, they share a similar distribution on the five frequent forms of triads (as plotted in Figure 2.11).

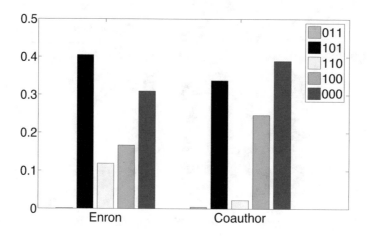

Figure 2.11: Social status. Distribution of five most frequent formations of triads with social status. Given a triad (A, B, C), let us use 1 to denote the advisor-advisee relationship and 0 colleague relationship. Thus, the number 011 to denote A and B are colleagues, B is C's advisor and A is C's advisor.

Opinion Leader. The two-step flow theory [89] suggests that ideas (innovations) usually flow first to *opinion leaders*, and then from them to a wider population. Our basic idea here is to examine whether "opinion leaders" are more likely to have a higher social status (manager or advisor) than ordinary users. To do this, we first categorize users into two groups (opinion leaders and ordinary users) by PageRank.[15] With PageRank, according to the network structure, we select as opinion leaders the top 1% users who have the highest PageRank scores and the rest as ordinary users. Then, we examine the probabilities that two users (A and B) have a directed social relationship (from higher social-status user to lower social-status user) such as advisor-advisee relationship or manager-subordinate relationship. Figure 2.12 shows some interesting discoveries. First, in both of the Enron and Coauthor networks, opinion leaders (detected by PageRank) are more likely (+71%-+84%) to have a higher social status than ordinary users. Second, and also more interestingly, in Enron, it is likely that ordinary users have a higher social status than opinion leaders. Its average likelihood is much larger (30 times) than that in the Coauthor network. The reason might be in the enterprise email network (Enron), some managers may be inactive, and most management-related communications were done by their assistants.

Summary. According to the statistics above, we have the following intuitions.

1. Probabilities of balanced triads based on communication links are very different in different networks, while the balance probabilities based on friendships (or trustful relationships) are similar with each other.

[15]PageRank is an algorithm to estimate the importance of each node in a network [116].

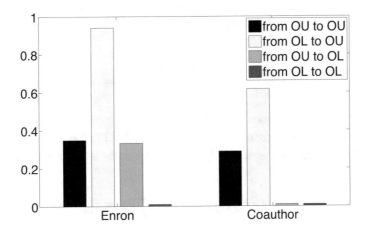

Figure 2.12: Opinion leader. OL - Opinion leader; OU - Ordinary user. Probability that two types of users have a directed relationship (from higher social status to lower status, i.e., manager-subordinate relationship in Enron and advisor-advisee relationship in Coauthor). It is clear that opinion leaders (detected by PageRank) are more likely to have a higher social status than ordinary users.

2. Users are more likely (+25%−+152% higher than chance) to have the same type of relationship with a user who spans a structural hole.

3. Most triads (99%) satisfy properties of the social status theory. For the five most frequent formations of triads, the Coauthor and the Enron networks share a similar distribution.

4. Opinion leaders are more likely (+71%−+84% higher than chance) to have a higher social status than ordinary users.

Model Framework

We propose a transfer-based factor graph (TranFG) model for learning and predicting the type of social relationships across networks. We first describe the model in the context of a single network, and then explain how to transfer the supervised information provided by one network to another network.

Learning over Single Network. Given a network $G = (V, E^L, E^U, \mathbf{X})$, each relationship (edge) e_i is associated with an attribute vector $\mathbf{x}_i$ and a label y_i indicates the type of the relationship. Let $\mathbf{X} = \{\mathbf{x}_i\}$ and $Y = \{y_i\}$. Then we have the following formulation:

$$P(Y|\mathbf{X}, G) = \frac{P(\mathbf{X}, G|Y)P(Y)}{P(\mathbf{X}, G)}. \tag{2.42}$$

Here, G denotes all forms of network information. This probabilistic formulation indicates that labels of edges depend on not only local attributes associated with each edge, but also the structure of the network. According to Bayes' rule, we have

$$P(Y|\mathbf{X},G) = \frac{P(\mathbf{X},G|Y)P(Y)}{P(\mathbf{X},G)}, \propto P(\mathbf{X}|Y) \cdot P(Y|G) \qquad (2.43)$$

where $P(Y|G)$ represents the probability of labels given the structure of the network and $P(\mathbf{X}|Y)$ denotes the probability of generating attributes $\mathbf{X}$ associated to all edges given their labels Y. We assume that the generative probability of attributes given the label of each edge is conditionally independent, thus we have

$$P(Y|\mathbf{X},G) \propto P(Y|G) \prod_i P(\mathbf{x}_i|y_i) \qquad (2.44)$$

where $P(\mathbf{x}_i|y_i)$ is the probability of generating attributes $\mathbf{x}_i$ given the label y_i. Now, the problem is how to instantiate the probability $P(Y|G)$ and $P(\mathbf{x}_i|y_i)$. In principle, they can be instantiated in different ways, for example by the Bayesian theory or Markov random fields. In this book, we choose the latter. Based on Markov random fields, for any node v_i, the conditional property holds: $P(y_i|G\backslash v_i) = P(y_i|NB(i))$, where $NB(i)$ are neighborhood of y_i in the graph G. The Hammersley-Clifford theorem [63] tells us that the probability of a Markov random field is equivalent to a Gibbs distribution which is factorized into positive function defined on cliques $\{Y_c\}$ that cover all the nodes and edges of G. Thus, the two probabilities in Eq. (2.44) can be rewritten as:

$$P(\mathbf{x}_i|y_i) = \frac{1}{Z_1} \exp\{\sum_{j=1}^{d} \alpha_j g_j(x_{ij}, y_i)\} \qquad (2.45)$$

$$P(Y|G) = \frac{1}{Z_2} \exp\{\sum_c \sum_k \mu_k h_k(Y_c)\} \qquad (2.46)$$

where Z_1 and Z_2 are normalization factors. Equation (2.45) indicates that we define a feature function $g_j(x_{ij}, y_i)$ for each attribute x_{ij} associated with edge e_i and α_j is the weight of the jth attribute. It can be defined as either a binary function or a real-valued function. For example, for inferring advisor-advisee relationships from the publication network, we can define a real-valued feature function as the difference of years when authors v_i and v_j, respectively, published his first paper. In Eq. (2.46), we define a set of correlation feature functions $\{h_k(Y_c)\}_k$ over each clique Y_c in the network. Here μ_k is the weight of the kth correlation feature function. The simplest clique is an edge, thus a feature function $h_k(y_i, y_j)$ can be defined as the correlation between two edges (e_i, e_j), if the two edges share a common end node. We also consider triads as cliques in the TranFG model, in that several social theories we discussed in Section 2.3.5 are based on triads.

If we are given a single network G with labeled information Y, learning the predictive model is to estimate a parameter configuration $\theta = (\{\alpha\}, \{\mu\})$ to maximize the log-likelihood objective function $\mathcal{O}(\theta) = \log P_\theta(Y|\mathbf{X}, G)$, i.e.,

$$\theta^\star = \arg \max \mathcal{O}(\theta). \qquad (2.47)$$

Learning across Heterogeneous Networks. We now turn to discuss how to learn the predictive model with two heterogeneous networks (a source network G_S and a target network G_T). Straightforwardly, we can define two separate objective functions for the two networks. The challenge is then how to bridge the two networks, so that we can transfer the labeled information from the source network to the target network. As the source and target networks may be from arbitrary domains, it is difficult to define correlations between them based on prior knowledge.

To this end, we propose a transfer-based factor graph (TranFG) model. Our idea is based on the fact that the social theories we discussed in Section 2.3.5 are general over all networks. Intuitively, we can leverage the correlation to the extent to which different networks satisfy each of the social theories to transfer the knowledge across networks. In particular, for social balance, we define triad-based features to denote the proportion of different balanced triangles in a network; for structural hole, we define edge correlation based features, i.e., correlation between two relationships e_i and e_j; for social status, we define features over triads to, respectively, represent the probabilities of the seven most frequent formations of triads; for opinion leaders, we define features over each edge.

Finally, by incorporating the social theories into our predictive model, we define the following log-likelihood objective function over the source and the target networks:

$$\begin{aligned}
\mathcal{O}(\alpha, \beta, \mu) &= \mathcal{O}_S(\alpha, \mu) + \mathcal{O}_T(\beta, \mu) \\
&= \sum_{i=1}^{|V_S|} \sum_{j=1}^{d} \alpha_j g_j(x_{ij}^S, y_i^S) + \sum_{i=1}^{|V_T|} \sum_{j=1}^{d'} \beta_j g_j'(x_{ij}^T, y_i^T) \\
&\quad + \sum_k \mu_k \Big(\sum_{c \in G_S} h_k(Y_c^S) + \sum_{c \in G_T} h_k(Y_c^T) \Big), \\
&\quad - \log Z
\end{aligned} \qquad (2.48)$$

where d and d' are numbers of attributes in the source network and the target network, respectively. In this objective function, the first term and the second term, respectively, define the likelihood over the source network and the target network; while the third term defines the likelihood over common features defined in the two networks. The common feature functions are defined according to the social theories. Such a definition implies that attributes of the two networks can be entirely different as they are optimized with different parameters $\{\alpha\}$ and $\{\beta\}$, while the information transferred from the source network to the target network is the importance of common features that are defined according to the social theories. Finally, we define four (real-

---

Input: a source network G_S, a target network G_T, and the learning rate η
Output: estimated parameters $\theta = (\{\alpha\}, \{\beta\}, \{\mu\})$

Initialize $\boldsymbol{\theta} \leftarrow 0$;
Perform statistics according to social theories;
Construct social theories based features $h_k(Y_c)$;
repeat

> **Step 1**: Perform LBP to calculate marginal distribution of unknown variables in the source network $P(y_i|x_i, G_S)$;
> **Step 2**: Perform LBP to calculate marginal distribution of unknown variables in the target network $P(y_i|x_i, G_T)$;
> **Step 3**: Perform LBP to calculate the marginal distribution of clique c, i.e., $P(y_c|\mathbf{X}_c^S, \mathbf{X}_c^T, G_S, G_T)$;
> **Step 4**: Calculate the gradient of μ_k according to Eq. (2.49) (for α_j and β_j with a similar formula);
> **Step 5**: Update parameter θ with the learning rate η:
>
> $$\boldsymbol{\theta}_{\text{new}} = \boldsymbol{\theta}_{\text{old}} + \eta \cdot \frac{\mathcal{O}(\theta)}{\theta}$$

until *Convergence*;

Algorithm 3: Learning algorithm for TranFG.

valued) balance based features, seven (real-valued) status based features, four (binary) features for opinion leader, and six (real-valued) correlation features for structural hole. More details about feature function are given in the Appendix.

Model Learning and Inferring. The last issue is to learn the TranFG model and to infer the type of unknown relationships in the target network. Learning the TranFG model is to estimate a parameter configuration $\theta = (\{\alpha\}, \{\beta\}, \{\mu\})$ to maximize the log-likelihood objective function $\mathcal{O}(\alpha, \beta, \mu)$. We use a gradient decent method (or a Newton-Raphson method) to solve the objective function. We use μ as the example to explain how we learn the parameters. Specifically, we first write the gradient of each μ_k with regard to the objective function:

$$
\begin{aligned}
\frac{\mathcal{O}(\theta)}{\mu_k} &= \mathbb{E}[h_k(Y_c^S) + h_k(Y_c^T)] \\
&\quad - \mathbb{E}_{P_{\mu_k}(Y_c|X_S, X_T, G_S, G_T)}[h_k(Y_c^S) + h_k(Y_c^T)],
\end{aligned}
\tag{2.49}
$$

where $\mathbb{E}[h_k(Y_c^S) + h_k(Y_c^T)]$ is the expectation of factor function $h_k(Y_c^S) + h_k(Y_c^T)$ given the data distribution (i.e., the average value of the factor function $h_k(Y_c)$ over all triads in the source and the target networks); the second term $\mathbb{E}_{P_{\mu_k}(Y_c|X_S, X_T, G_S, G_T)}[.]$ is the expectation under the distribution $P_{\mu_k}(Y_c|\mathbf{X}_S, \mathbf{X}_T, G_S, G_T)$ given by the estimated model. Similar gradients can be derived for parameters α_j and β_j.

As the graphical structure can be arbitrary and may contain cycles, we use loopy belief propagation (LBP) [111] to approximate the gradients. It is worth noting that in order to leverage the unlabeled relationships, we need to perform the LBP process twice in each iteration, one time for estimating the marginal distribution of unknown variables $y_i = ?$ and the other time for marginal distribution over all cliques. Finally, with the gradient we update each parameter with a learning rate η. The learning algorithm is summarized in Algorithm 3. We see that in the learning process, the algorithm uses an additional loopy belief propagation to infer the label of unknown relationships. After learning, all unknown relationships are assigned with labels that maximize the marginal probabilities.

Evaluation

The proposed framework is very general and can be applied to many different networks. For experiments, we consider five different types of networks: Epinions, Slashdot, Mobile, Coauthor, and Enron. On the first three networks (Epinions, Slashdot, and Mobile), our goal is to infer undirected relationships (e.g., friendships), while on the other two networks (Coauthor and Enron), the goal is to infer directed relationships (e.g., advisor-advisee relationships).

Evaluation Measures. To quantitatively evaluate the performance of inferring the type of social relationships, we conducted experiments with different pairs of (source and target) networks, and evaluated the proposed approaches in terms of Precision, Recall, and F1-Measure. We compare the following methods for inferring the type of social relationships.

SVM: Similar to the logistic regression model used in Leskovec et al. [92], SVM uses attributes associated with each edge as features to train a classification model and then employs the classification model to predict edges' labels in the test data set. For SVM, we employ SVM-light.

CRF: It trains a conditional random field [87] with attributes associated with each edge and correlations between edges.

PFG: The method is also based on CRF, but it employs the unlabeled data to help learn the predictive model. The method is proposed in Tang et al. [150].

TranFG: The proposed approach, which leverages the label information from the source network to help infer the type of relationship in the target network.

We also compare with the method TPFG proposed in Wang et al. [155] for mining advisor-advisee relationships in the publication network. This method is domain-specific and thus we only compare with it on the Coauthor network.

In all experiments, we use the same feature definitions for all methods. On the Coauthor network, we do not consider some domain-specific correlation features.[16] All codes were implemented in C++, and all experiments were performed on a PC running Windows 7 with Intel (R) Core (TM) 2 CPU 6600 (2.4 GHz) and 4 GB memory. It took about 1–30 min to train the TranFG model over different data sets (e.g., 30 min for learning over the Epinions and the Slash-

---

[16]We conducted experiments, but found that those features will lead to overfitting.

dot networks). For incorporating social balance and social status into the TranFG model, we need count all triads in the source and the target networks. We design an efficient linear algorithm[1], which takes 1–5 min to enumerate all triads for the five networks.

Table 2.11: Performance comparison of different methods for inferring friendships (or trustful relationships). (S) indicates the source network and (T) the target network. For the target network, we use 40% of the labeled data in training and the rest for test

Data Set	Method	Prec.	Rec.	F1-score
Epinions (S) to Slashdot (T) (40%)	SVM	0.7157	**0.9733**	0.8249
	CRF	0.8919	0.6710	0.7658
	PFG	0.9300	0.6436	0.7607
	TranFG	**0.9414**	0.9446	**0.9430**
Slashdot (S) to Epinions (T) (40%)	SVM	0.9132	**0.9925**	0.9512
	CRF	0.8923	0.9911	0.9393
	PFG	**0.9954**	0.9787	**0.9870**
	TranFG	**0.9954**	0.9787	**0.9870**
Epinions (S) to Mobile (T) (40%)	SVM	0.8983	0.5955	0.7162
	CRF	0.9455	0.5417	0.6887
	PFG	**1.0000**	0.5924	0.7440
	TranFG	0.8239	**0.8344**	**0.8291**
Slashdot (S) to Mobile (T) (40%)	SVM	0.8983	0.5955	0.7162
	CRF	0.9455	0.5417	0.6887
	PFG	**1.0000**	0.5924	0.7440
	TranFG	0.7258	**0.8599**	**0.7872**

Inferring Accuracy Analysis. We compare the performance of the four methods for inferring friendships (or trustful relationships) on four pairs of networks: Epinions (S) to Slashdot (T), Slashdot (S) to Epinions (T), Epinions (S) to Mobile (T), and Slashdot (S) to Mobile (T).[17] In all experiments, we use 40% of the labeled data in the target network for training and the rest for test. For transfer, we consider the labeled information in the source network. Table 2.11 lists the performance of the different methods on the four test cases. Our approach shows better performance than the three alternative methods. We conducted sign tests for each result, which shows that all the improvements of our approach TranFG over the three methods are statistically significant ($p \ll 0.01$).

Table 2.12 shows the performance of the four methods for inferring directed relationships (the source end has a higher social status than the target end) on two pairs of networks: Coauthor

---

[17]We did try to use Mobile as the source network and Slashdot/Epinions as the target network. However, as the size of Mobile is much smaller than the other two networks, the performance was considerably worse.

Table 2.12: Performance comparison of different methods for inferring directed relationships (the source end has a higher social status than the target end). (S) indicates the source network and (T) the target network. For the target network, we use 40% of labeled data in training and the rest for test

Data Set	Method	Prec.	Rec.	F1-score
Coauthor (S) to Enron (T) (40%)	SVM	0.9524	0.5556	0.7018
	CRF	0.9565	0.5366	0.6875
	PFG	**0.9730**	0.6545	0.7826
	TranFG	0.9556	**0.7818**	**0.8600**
Enron (S) to Coauthor (T) (40%)	SVM	0.6910	0.3727	0.4842
	CRF	**1.0000**	0.3043	0.4666
	PFG	0.9916	0.4591	0.6277
	TPFG	0.5936	**0.7611**	0.6669
	TranFG	0.9793	0.5525	**0.7065**

(S) to Enron (T) and Enron (S) to Coauthor (T). We use the same experimental setting as that for inferring friendships on the four pairs of networks, i.e., taking 40% of the labeled data in the target network for training and the rest for test, while for transfer, analogously, we consider the labeled information from the source network. We see that by leveraging the supervised information from the source network, our method clearly improves the performance (about 15% by F1-score on Enron and 10% on Coauthor).

The method PFG can be viewed as a non-transferable counterpart of our method, which does not consider the labeled information from the source network. From both Table 2.11 and Table 2.12, we can see that with the transferred information, our method can clearly improve the relationship categorization performance. Another phenomenon is that PFG has a better performance than the other two methods (SVM and CRF) in most cases. PFG could leverage the unlabeled information in the target network, thus enhances the inferring performance. The only exception is the case of Epinions (S) to Slashdot (T), where it seems that users in Slashdot have a relatively consistent pattern, thus a classification based method (SVM) with only general features (e.g., in-degree, out-degree, and number of common neighbors) can achieve very high performance.

Factor Contribution Analysis. We now analyze how different social theories (social balance, social status, structural hole, and two-step flow (opinion leader)) can help infer social ties. For inferring friendships, we consider social balance-(SB) and structural hole-(SH) based transfer and for inferring directed friendships, we consider social status-(SS) and opinion leader-(OL) based transfer. Here we examine the contribution of the different factors defined in our TranFG model. Figure 2.13 shows the average F1-Measure score over the different networks, obtained by the TranFG model for inferring friendships and directed relationships. In particular, TranFG-SB

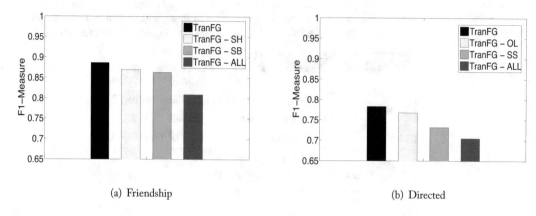

(a) Friendship   (b) Directed

Figure 2.13: Factor contribution analysis. TranFG-SH denotes our TranFG model by ignoring the structural hole-based transfer. TranFG-SB stands for ignoring the structural balance-based transfer. TranFG-OL stands for ignoring the opinion leader-based transfer and TranFG-SS stands for ignoring social status-based transfer.

represents TranFG without social balance based features and TranFG-All denotes that we remove all the transfer features. It can be clearly observed that the performance drops when ignoring each of the factors. We can also see that for inferring friendships the social balance is a bit more useful than structural hole, and for inferring directed relationships the social status factor is more important than the factor of opinion leader. The analysis also confirms that our method works well (further improvement is obtained) when combining different social theories.

Social Balance and Structural Hole Based Transfer. We present an in-depth analysis on how the social balance and structural hole based transfer can help by varying the percent of labeled training data in the target network. We see that in all cases except Slashdot-to-Epinions, clear improvements can be obtained by using the social balance- and structural hole-based transfer, when the labeled data in the target network is limited ($\leq$ 50%). Indeed, in some cases such as Epinions-to-Slashdot, with merely 10% of the labeled relationships in Slashdot, our method can obtain a good performance (88% by F1-score). Without transfer, the best performance is only 70% (obtained by SVM). We also find that structural balance-based transfer is more helpful than structural hole-based transfer for inferring friendships in most cases with various percents of labeled relationships. This result is consistent with what we obtained in the factor contribution analysis.

   A different phenomenon is found in the case of Slashdot-to-Epinions, where all methods can obtain a F1-score of 94% with only 10% of the labeled data. The knowledge transfer seems not helpful. By a careful investigation, we found simply with those features (cf. Appendix for

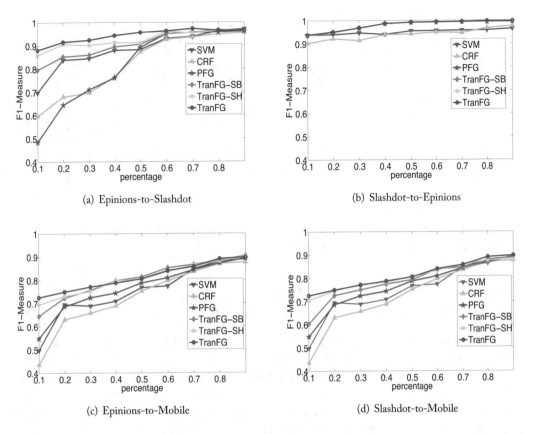

Figure 2.14: Performance of inferring friendships with and without the balance-based transfer by varying the percent of labeled data in the target network.

details) defined on the edges, we could achieve a high performance (about 90%). The structure information indeed helps, but the gained improvement is limited.

Social Status and Opinion Leader Based Transfer. Figure 2.15 shows an analysis for inferring directed relationships on the two cases (Enron-to-Coauthor and Coauthor-to-Enron). Here, we focus on testing how social status and opinion leader based transfer can help infer the type of relationships by varying the percent of labeled relationships in the target network. In both cases (Coauthor-to-Enron and Enron-to-Coauthor), the TranFG model achieves consistent improvements. For example, when there is only 10% of labeled advisor-advisee relationships in the Coauthor network, without considering the status and opinion leader based transfer, the F1-score is only 24%. By leveraging the status and opinion leader-based transfer from the email network

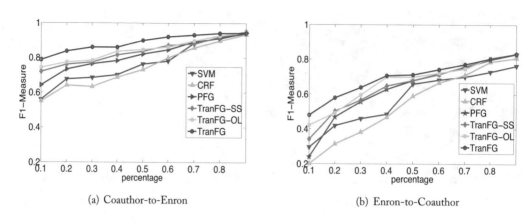

(a) Coauthor-to-Enron

(b) Enron-to-Coauthor

Figure 2.15: Performance of inferring directed relationship with and without the status-based transfer by varying the percent of labeled data in the target network.

(Enron), the score is doubled (47%). Moreover, we find that the social status-based transfer is more helpful than the opinion leader-based transfer with various percents of the labeled data.

Qualitative Case Study. Now we present a case study to demonstrate the effectiveness of the proposed model. Figure 2.16 shows an example generated from our experiments. It represents a portion of the Coauthor network. Black edges and arrows, respectively, denote labeled colleague relationships and advisor-advisee relationships in the training data. Colored arrows and edges indicate advisor-advisee and colleagues relationships detected by three methods: SVM, PFG, and TranFG, with red color indicating mistake ones. The numbers associated with each author, respectively, denote the number of papers and the score of h-index.

We investigate more by looking at a specific example. SVM mistakenly classifies three advisor-advisee relationships and two colleague relationships. SVM trains a local classification model without considering the network information. PFG considers the network information as well as the unlabeled data, thus obtains a better result. Our proposed TranFG model further corrects two mistakes ("Fait-Leonardi" and "Ausiello-Laura") by leveraging properties of social status and opinion leader. For example, the results obtained by PFG among "Azar," "Amos," and "Leonardi" form a triad of ("011"). Although it satisfies the property of social status, the probability of such triad is much lower (0.4% vs. 24.6%) than the form ("100"). However, the limitation of the training data leads PFG to result in a bias mistake (5.8% vs. 12.6%). TranFG smoothes the results by transferring knowledge from the source (Enron) network.

Summary

In this section, we study the problem of inferring social ties across heterogeneous networks. We precisely define the problem and propose a transfer-based factor graph (TranFG) model. The

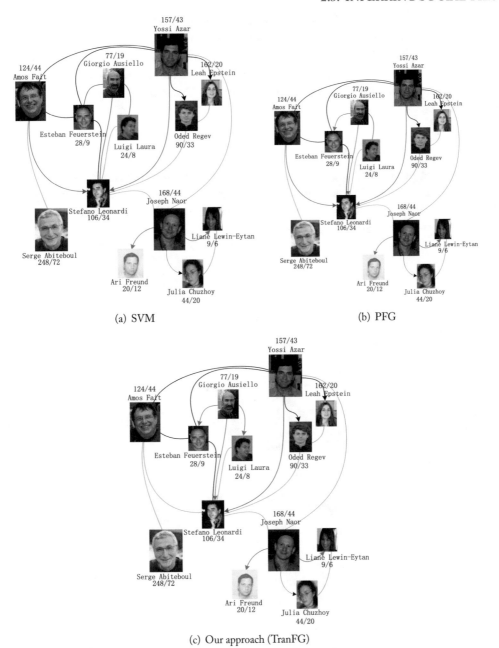

Figure 2.16: Case study. Illustration of inferring advisor-advisee relationships on the Coauthor network. Directed edges indicate advisor relationships, and undirected ones indicate co-author relationships. Black edges indicate labeled data. Red colored edges indicate wrong predictions.

model incorporates social theories into a semi-supervised learning framework, which is used to transfer supervised information from the source network to help infer social ties in the target network. We evaluate the proposed model on five different genres of networks. We show that the proposed model can significantly improve the performance for inferring social ties across different networks comparing with several alternative methods. Our study also reveals several interesting phenomena.

2.4 CONCLUSIONS

In this chapter, we give a comprehensive introduction to the study of social tie analysis. We present the state-of-the-art algorithms for predicting missing links and describe methodologies for inferring social ties including unsupervised learning-based method, supervised learning-based method, active learning-based method, and transfer learning-based method.

CHAPTER 3

Social Influence Analysis

Social influence occurs when one's opinions, emotions, or behaviors are affected by others, intentionally or unintentionally [78]. Social influence is a prevalent, complex, and subtle force that governs the dynamics of all social networks.

3.1 OVERVIEW

Social influence has been a widely accepted phenomenon in social networks for decades. Many applications have been built based around the implicit notation of social influence between people. For example, more and more people make decisions based on their interactions from social networks. People often pick what restaurants to go to based on recommendations and reviews from Yelp. As the use of social networks grows in all domains, such behaviors of social influence become more and more prevalent. More and more people make decisions and changes influenced by their social networks. With the exponential growth of online social network services such as Facebook and Twitter, social influence can for the first time be measured over a large population.

Deutsch and Gerard [37] categorized social influence into informative social influence and normative social influence from the perspective of psychological needs. The former is an influence to accept (or disagree with) information from others and the latter is an influence to conform to the expectations of others. With the power of influence, a company can market a new product by first convincing a small number of influential users to adopt the product and then triggering a cascade of further adoptions through the effect of "word of mouth" in the social network. In an academic network, with the influence between research collaborators, novel ideas or innovations can quickly spread and lead to the blooming of new academic directions. Christakis and Fowler [49] created the theory of Three-Degree-of-Influence, which posits that "everything we do or say tends to ripple through our network, having an impact even on our three degree of friends (friends' friends' friends)."

Recently, social influence analysis has attracted considerable research interest. Roughly speaking, existing works can be grouped into three categories. The first category of research focused on qualitatively validating the existence of influence [2, 9, 18, 31]. For example, Bond et al. [18] used a randomized controlled trial to verify the social influence on political voting behaviors by delivering political mobilization messages to 61 million Facebook users during the 2010 US congressional elections. They found that the messages directly influence political self-expression and real-world voting behavior of millions of people. Bakshy et al. [9] also conducted two very large field experiments on Facebook to test the effect of social influence on consumer responses to

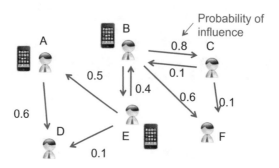

Figure 3.1: Influence maximization for viral marketing.

advertising. They also found significant influence exists for users' advertising behaviors, and the the greatest influence occurs with strong ties. The second category of research on social influence is to measure the influential strength between users. Tang et al. [143] first studied the problem of topic-level social influence analysis and presented a *Topical Affinity Propagation* (TAP) approach to quantify the social influence between users from different angles. Goyal et al. [55] proposed a method to learn the influence probability by considering the correlation between users' actions. Liu et al. [99] extended the influence learning to heterogeneous social networks and further studied the influence propagation and aggregation mechanisms. The third category of social influence research is social influence model. Two popular social influence models are *linear threshold model* and *independent cascaded model* [42, 79, 121]. In both models, the objective is to find a small subset of users (seed users) to adopt a behavior (e.g., adopt a product), and the goal is to trigger a large cascade of further adoptions through the influence diffusion model. The problem is referred to as influence maximization. Richardson [121] and Kempe et al. [79] formally defined the problem of influence maximization. Chen et al. [27] presented an efficient algorithm to solve the problem. Goyal et al. [54] leveraged real propagation traces to derive more accurate influence maximization models.

Figure 3.1 shows an example of using social influence for viral marketing. A mobile phone company wants to advertise their new mobile phone in the social network. Their strategy is to find a small number of influential (seed) users in a social network to freely use the new product. The objective is that the adoption of the small number of influential users can trigger their friends and further continue to trigger their friends' friends to also use the mobile phone. The real number associated with each arrow indicates the influence of the source user on the target user. From the example, we see that there are several challenges for finding the optimal subset of seed users. The first one is how to obtain (or quantify) the influence between users and the second one is how to design an algorithm to select the seed users.

In addition, the effect of the social influence from different angles (topics) may be different. For example, in the research community, such influences are well known. Most researchers are influenced by others in terms of collaboration and citations. An expert in "data mining" would have a strong influence on his collaborators, while on some other topic, such as "computer graphics," he may be mainly influenced by his collaborators. Thus, the key question is how to effectively and efficiently quantify the influence among users on different topics.

In this chapter, we will focus on introducing methodologies for quantifying the topic-level influential strength between users for large social networks, and then introduce its application to social action prediction.

3.2   MINING TOPIC-LEVEL SOCIAL INFLUENCE ANALYSIS

The goal of social influence analysis is to derive the topic-level social influences based on the input network and topic distribution on each node. First we introduce some terminology, and then define the social influence analysis problem.

Topic Distribution. In social networks, a user usually has interests on multiple topics. Formally, each node $v \in V$ is associated with a vector $\theta_v \in \mathbb{R}^T$ of T-dimensional topic distribution ($\sum_z \theta_{vz} = 1$). Each element θ_{vz} is the probability(importance) of the node on topic z.

Topic-Based Social Influences. Social influence from node s to t denoted as μ_{st} is a numerical weight associated with the edge e_{st}. In most cases, the social influence score is asymmetric, i.e., $\mu_{st} \neq \mu_{ts}$. Furthermore, the social influence from node s to t will vary on different topics.

Thus, based on the above concepts, we can define the tasks of topic-based social influence analysis. Given a social network $G = (V, E)$ and a topic distribution for each node, the goal is to find the topic-level influence scores on each edge.

Problem 3.1   Given (1) a network $G = (V, E)$, where V is the set of nodes (users, entities) and E is the set of directed/undirected edges, and (2) T-dimensional topic distribution $\theta_v \in \mathbb{R}^T$ for all node v in V, how does one find the topic-level influence network $G_z = (V_z, E_z)$ for all topics $1 \leq z \leq T$? Here V_z is a subset of nodes that are related to topic z and E_z is the set of pair-wise weighted influence relations over V_z, each edge is the form of a triplet (v_s, v_t, μ_{st}^z) (or shortly (e_{st}, μ_{st}^z)), where the edge is from node v_s to node v_t with the weight μ_{st}^z.

The input to our social influence analysis includes: (1) networks and (2) topic distribution on all nodes. The first input is the network backbone obtained by any social network, such as online social networks like Facebook and Twitter. The second input is the topic distribution for all nodes. In general, the topic information can be obtained in many different ways. For example, in a social network, one can use the predefined categories as the topic information, or use user-assigned tags as the topic information. In addition, we can use statistical topic modeling [14, 69] to automatically extract topics from the social networking data. In this definition, to make it general, we do not consider users' actions. We will extend the problem formulation in the next section.

The social influence analysis problem poses a unique set of challenges.

First, how does one leverage both node-specific topic distribution and network structure to quantify social influence? In another words, a user's influence on others not only depends on their own topic distribution, but also relies on what kinds of social relationships they have with others. The goal is to design a unified approach to utilize both the local attributes (topic distribution) and the global structure (network information) for social influence analysis.

Second, how does one scale the proposed analysis to a real large social network? For example, the academic community of Computer Science has more than 1 million researchers and more than 10 million coauthor relations; Facebook has more than 50 millions users and hundreds of millions of different social ties. How to efficiently identify the topic-based influential strength for each social tie is really a challenging problem.

Solution. To address the above challenges, we propose Topical Affinity Propagation (TAP) to model the topic-level social influence on large networks. In particular, given a social network $G = (V, E)$ and a topic model on the nodes V, TAP computes topic-level social influence graphs $G_z = (V_z, E_z)$ for all topic $1 \leq z \leq T$. The key features of TAP are the following:

- TAP provides topical influence graphs that quantitatively measure the influence on a fine-grain level;

- the influence graphs from TAP can be used to support other applications such as finding representative nodes or constructing the influential subgraphs; and

- an efficient distributed learning algorithm is developed for TAP based on the Map-Reduce framework in order to scale to real large networks.

3.2.1   TOPICAL AFFINITY PROPAGATION

Based on the input network and topic distribution on the nodes, we formalize the social influence problem in a topical factor graph model and propose a topical affinity propagation on the factor graph to automatically identify the topic-specific social influence. Our main idea is to leverage an affinity propagation at the topic-level for social influence identification. The approach is based on the theory of factor graph [84], in which the observation data are cohesive on both local attributes and relationships. In our setting, the node corresponds to the observation data in the factor graph and the social relationship corresponds to edge between the observed nodes and the observation data in the graph. Finally, we propose two different propagation rules: one based on message passing on graphical models, the other a parallel update rule that is suitable for Map-Reduce framework.

Topical Factor Graph (TFG) model. Now we first explain the proposed TFG model. The TFG model has the following components: a set of observed variables $\{v_i\}_{i=1}^{N}$ and a set of hidden vectors $\{y_i\}_{i=1}^{N}$, which corresponds to the N nodes in the input network. Notations are summarized in Table 3.1.

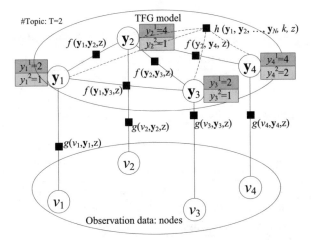

Figure 3.2: Graphical representation of the topical factor graph model. $\{v_1, \ldots, v_4\}$ are observable nodes in the social network; $\{\mathbf{y}_1, \ldots, \mathbf{y}_4\}$ are hidden vectors defined on all nodes, with each element representing which node has the highest probability to influence the corresponding node; $g(.)$ represents a feature function defined on a node, $f(.)$ represents a feature function defined on an edge; and $h(.)$ represents a global feature function defined for each node, i.e., $k \in \{1, \ldots, N\}$.

The hidden vector $\mathbf{y}_i \in \{1, \ldots, N\}^T$ models the topic-level influences from other nodes to node v_i. Each element y_i^z, taking the value from the set $\{1, \ldots, N\}$, represents the node that has the highest probability to influence node v_i on topic z.

For example, Figure 3.2 shows a simple example of an TFG. The observed data consists of four nodes $\{v_1, \ldots, v_4\}$, which have corresponding hidden vectors $\mathbf{Y} = \{\mathbf{y}_1, \ldots, \mathbf{y}_4\}$. The edges between the hidden nodes indicate the four social relationships in the original network (aka the edges of the input network).

There are three kinds of feature functions:

- **Node feature function** $g(v_i, \mathbf{y}_i, z)$ is a feature function defined on node v_i specific to topic z;

- **Edge feature function** $f(\mathbf{y}_i, \mathbf{y}_j, z)$ is a feature function defined on the edge of the input network specific to topic z; and

- **Global feature function** $h(\mathbf{y}_1, \ldots, \mathbf{y}_N, k, z)$ is a feature function defined on all nodes of the input network w.r.t. topic z.

Basically, node feature function g describes local information on nodes, edge feature function f describes dependencies between nodes via the edge on the graph model, and global feature function captures constraints defined on the network.

Table 3.1: Notations

SYMBOL	DESCRIPTION
N	number of nodes in the social network
M	number of edges in the social network
T	number of topics
V	the set of nodes in the social network
E	the set of edges
v_i	a single node
y_i^z	node-v_i's representative on topic z
$\mathbf{y}_i$	the hidden vector of representatives for all topics on node v_i
θ_i^z	the probability for topic z to be generated by the node v_i
e_{st}	an edge connecting node v_s and node v_t
w_{st}^z	the similarity weight of the edge e_{st} w.r.t. topic z
μ_{st}^z	the social influence of node v_s on node v_t w.r.t. topic z

In this book, we define the node feature function g as:

$$
g(v_i, \mathbf{y}_i, z) = \begin{cases} \dfrac{w_{iy_i^z}^z}{\sum_{j \in NB(i)}(w_{ij}^z + w_{ji}^z)} & y_i^z \neq i \\[3mm] \dfrac{\sum_{j \in NB(i)} w_{ji}^z}{\sum_{j \in NB(i)}(w_{ij}^z + w_{ji}^z)} & y_i^z = i \end{cases}, \tag{3.1}
$$

where $NB(i)$ represents the indices of the neighboring nodes of node v_i; $w_{ij}^z = \theta_j^z \alpha_{ij}$ reflects the topical similarity or interaction strength between v_i and v_j, with θ_j^z denoting the importance of node-j to topic z, and α_{ij} denoting the weight of the edge e_{ij}. α_{ij} can be defined by different ways. For example, in a coauthor network, α_{ij} can be defined as the number of papers coauthored by v_i and v_j. The above definition of the node feature function has the following intuition: if node v_i has a high similarity/weight with node v_{y_i}, then v_{y_i} may have a high influence on node v_i; or if node v_i is trusted by other users, i.e., other users take him as an high influential node on them, then it must also "trust" himself highly (taking himself as a most influential user on him).

As for the edge feature function, we define a binary feature function, i.e., $f(\mathbf{y}_i, \mathbf{y}_j, z) = 1$ if and only if there is an edge e_{ij} between node v_i and node v_j, otherwise 0. We also define a global edge feature function h on all nodes, i.e.:

$$
h(\mathbf{y}_1, \ldots, \mathbf{y}_N, k, z) = \begin{cases} 0 & \text{if } y_k^z = k \text{ and } y_i^z \neq k \text{ for all } i \neq k \\ 1 & \text{otherwise.} \end{cases} \tag{3.2}
$$

Intuitively, $h(\cdot)$ constrains the model to bias towards the "true" representative nodes. More specially, a representative node on topic z must be the representative of itself on topic z, i.e., $y_k^z = k$. And it must be a representative of at least another node v_i, i.e., $\exists y_i^z = k, i \neq k$.

Next, a factor graph model is constructed based on this formulation. Typically, we hope that a model can best fit (reconstruct) the observation data, which is usually represented by maximizing the likelihood of the observation. Thus we can define the objective likelihood function as:

$$P(\mathbf{v}, \mathbf{Y}) = \frac{1}{Z} \prod_{k=1}^{N} \prod_{z=1}^{T} h(\mathbf{y}_1, \dots, \mathbf{y}_N, k, z)$$

$$\prod_{i=1}^{N} \prod_{z=1}^{T} g(v_i, \mathbf{y}_i, z) \prod_{e_{kl} \in E} \prod_{z=1}^{T} f(\mathbf{y}_k, \mathbf{y}_l, z), \tag{3.3}$$

where $\mathbf{v} = [v_1, \dots, v_N]$ and $\mathbf{Y} = [\mathbf{y}_1, \dots, \mathbf{y}_N]$ corresponds to all observed and hidden variables, respectively; g and f are the node and edge feature functions; h is the global feature function; Z is a normalizing factor.

The factor graph in Figure 3.2 describes this factorization. Each black box corresponds to a term in the factorization, and it is connected to the variables on which the term depends.

Based on this formulation, the task of social influence is cast as identifying which node has the highest probability to influence another node on a specific topic along with the edge. That is, to maximize the likelihood function $P(\mathbf{v}, \mathbf{Y})$. One parameter configuration is shown in Figure 3.2. On topic 1, both node v_1 and node v_3 are strongly influenced by node v_2, while node v_2 is mainly influenced by node v_4. On topic 2, the situation is different. Almost all nodes are influenced by node v_1, where node v_4 is indirectly influenced by node v_1 via the node v_2.

Basic TAP learning algorithm. To train the TFG model, we can take Equation (3.3) as the objective function to find the parameter configuration that maximizes the objective function. While it is intractable to find the exact solution to Equation (3.3), approximate inference algorithms such as sum-product algorithm [84], can be used to infer the variables $\mathbf{y}$.

In sum-product algorithm, messages are passed between nodes and functions. Message passing is initiated at the leaves. Each node v_i remains idle until messages have arrived on all but one of the edges incident on the node v_i. Once these messages have arrived, node v_i is able to compute a message to be sent onto the one remaining edge to its neighbor. After sending out a message, node v_i returns to the idle state, waiting for a "return message" to arrive from the edge. Once this message has arrived, the node is able to compute and send messages to each of neighborhood nodes. This process runs iteratively until convergence.

However, traditional sum-product algorithm cannot be directly applied for multiple topics. We first consider a basic extension of the sum-product algorithm: topical sum-product. The algorithm iteratively updates a vector of messages $\mathbf{m}$ between variable nodes and factor (i.e., feature function) nodes. Hence, two update rules can be defined, respectively, for a topic-specific message sent from variable node to factor node and for a topic-specific message sent from factor node to variable node:

$$m_{y \to f}(y,z) = \prod_{f' \sim y \setminus f} m_{f' \to y}(y,z) \prod_{z' \neq z} \prod_{f' \sim y \setminus f} m_{f' \to y}(y,z')^{(\tau_{z'z})}$$

$$m_{f \to y}(y,z) = \sum_{\sim \{y\}} \left(f(Y,z) \prod_{y' \sim f \setminus y} m_{y' \to f}(y',z) \right)$$

$$+ \sum_{z' \neq z} \tau_{z'z} \sum_{\sim \{y\}} \left(f(Y,z') \prod_{y' \sim f \setminus y} m_{y' \to f}(y',z') \right), \quad (3.4)$$

where

- $f' \sim y \setminus f$ represents f' is a neighbor node of variable y on the factor graph except factor f;

- Y is a subset of hidden variables that feature function f is defined on; for example, a feature $f(y_i, y_j)$ is defined on edge e_{ij}, then we have $Y = \{y_i, y_j\}$; $\sim \{y\}$ represents all variables in Y except y;

- the sum $\sum_{\sim \{y\}}$ actually corresponds to a marginal function for y on topic z; and

- coefficient τ represents the correlation between topics, which can be defined in many different ways. In this book, for simplicity, we assume that topics are independent. That is, $\tau_{zz'} = 1$ when $z = z'$ and $\tau_{zz'} = 0$ when $z \neq z'$. In the following, we will propose two new learning algorithms, which are also based this independent assumption.

New TAP learning algorithm. However, the sum-product algorithm requires that each node need wait for all(-but-one) message to arrive, thus the algorithm can only run in a sequential mode. This results in a high complexity of $O(N^4 \times T)$ in each iteration. To deal with this problem, we propose an affinity propagation algorithm, which converts the message passing rules into equivalent update rules passing message directly between nodes rather than on the factor graph. The algorithm is summarized in Algorithm 1. In the algorithm, we first use logarithm to transform sum-product into max-sum, and introduce two sets of variables $\{r_{ij}^z\}_{z=1}^T$ and $\{a_{ij}^z\}_{z=1}^T$ for each edge e_{ij}. The new update rules for the variables are as follows:

$$r_{ij}^z = b_{ij}^z - \max_{k \in NB(j)} \{b_{ik}^z + a_{ik}^z\} \quad (3.5)$$

$$a_{jj}^z = \max_{k \in NB(j)} \min \{r_{kj}^z, 0\} \quad (3.6)$$

$$a_{ij}^z = \min(\max\{r_{jj}^z, 0\}, -\min\{r_{jj}^z, 0\}$$

$$- \max_{k \in NB(j) \setminus \{i\}} \min\{r_{kj}^z, 0\}), i \in NB(j), \quad (3.7)$$

where $NB(j)$ denotes the neighboring nodes of node j, r_{ij}^z is the influence message sent from node i to node j and a_{ij}^z is the influence message sent from node j to node i, initiated by 0, and b_{ij}^z is the logarithm of the normalized feature function

$$b_{ij}^z = \log \frac{g(v_i, \mathbf{y}_i, z)|_{y_i^z = j}}{\sum_{k \in NB(i) \cup \{i\}} g(v_i, \mathbf{y}_i, z)|_{y_i^z = k}}. \tag{3.8}$$

The introduced variables r and a have the following nice explanation. Message a_{ij}^z reflects, from the perspective of node v_j, how likely node v_j thinks he/she influences on node v_i with respect to topic z, while message r_{ij}^z reflects, from the perspective of node v_i, how likely node v_i agrees that node v_j influence on him/her with respect to topic z. Finally, we can define the social influence score based on the two variables r and a using a sigmoid function:

$$\mu_{st}^z = \frac{1}{1 + e^{-(r_{ts}^z + a_{ts}^z)}}. \tag{3.9}$$

The score μ_{st}^z actually reflects the maximum of $P(\mathbf{v}, \mathbf{Y}, z)$ for $y_t^z = s$, thus the maximization of $P(\mathbf{v}, \mathbf{Y}, z)$ can be obtained by

$$y_t^z = \arg \max_{s \in NB(t) \cup \{t\}} \mu_{st}^z. \tag{3.10}$$

Finally, according to the obtained influence scores $\{\mu_{st}^z\}$ and the topic distribution $\{\theta_v\}$, we can easily generate the topic-level social influence graphs. Specifically, for each topic z, we first filter out irrelevant nodes, i.e., nodes that have a lower probability than a predefined threshold. An alternative way is to keep only a fixed number (e.g., 1,000) of nodes for each topic-based social influence graph. (This filtering process can be also taken as a preprocessing step of our approach, which is the way we conducted our experiments.) Then, for a pair of nodes (v_s, v_t) that has an edge in the original network G, we create two directed edges between the two nodes and assign the social influence scores μ_{st}^z and μ_{ts}^z, respectively. Finally, we obtain a directed social influence graph G_z for the topic z.

The new algorithm reduces the complexity of each iteration from $O(N^4 \times T)$ in the sum-product algorithm to $O(M \times T)$. More importantly, the new update rules can be easily parallelized.

Distributed TAP learning algorithm. As a social network may contain millions of users and hundreds of millions of social ties between users, it is impractical to learn a TFG from such a huge data using a single machine. To address this challenge, we deploy the learning task on a distributed system under the map-reduce programming model [36].

Map-Reduce is a programming model for distributed processing of large data sets. In the *map* stage, each machine (called a *process node*) receives a subset of data as input and produces a set of intermediate key/value pairs. In the *reduce* stage, each process node merges all intermediate values associated with the same intermediate key and outputs the final computation results. Users

Input: $G = (V, E)$ and topic distributions $\{\theta_v\}_{v \in V}$
Output: topic-level social influence graphs $\{G_z = (V_z, E_z)\}_{z=1}^T$

Calculate the node feature function $g(v_i, \mathbf{y}_i, z)$;
Calculate b_{ij}^z according to Equation (3.8);
Initialize all $\{r_{ij}^z\} \leftarrow 0$;
repeat
    foreach *edge-topic pair* (e_{ij}, z) **do**
        | Update r_{ij}^z according to Equation (3.5);
    end
    foreach *node-topic pair* (v_j, z) **do**
        | Update a_{jj}^z according to Equation (3.6);
    end
    foreach *edge-topic pair* (e_{ij}, z) **do**
        | Update a_{ij}^z according to Equation (3.7);
    end
until *convergence*;
foreach *node* v_t **do**
    foreach *neighboring node* $s \in NB(t) \cup \{t\}$ **do**
        | Compute μ_{st}^z according to Equation (3.9);
    end
end
Generate $G_z = (V_z, E_z)$ for every topic z according to $\{\mu_{st}^z\}$;

Algorithm 4: The new TAP learning algorithm.

specify a map function that processes a key/value pair to generate a set of intermediate key/value pairs, and a reduce function that merges all intermediate values associated with the same intermediate key.

In our affinity propagation process, we first partition the large social network graph into subgraphs and distribute each subgraph to a process node. In each subgraph, there are two kinds of nodes: internal nodes and marginal nodes. Internal nodes are those all of whose neighbors are inside the very subgraph; marginal nodes have neighbors in other subgraphs. For every subgraph G, all internal nodes and edges between them construct the closed graph $\bar{G}$. The marginal nodes can be viewed as "the supporting information" for updating the rules. For easy explanation, we consider the distributed learning algorithm on a single topic and thus the map stage and the reduce stage can be defined as follows.

In the map stage, each process node scans the closed graph $\bar{G}$ of the assigned subgraph G. Note that every edge e_{ij} has two values a_{ij}^z and r_{ij}. Thus, the map function is defined as for every key/value pair e_{ij}/a_{ij}, it issues an intermediate key/value pair $e_{i*}/(b_{ij} + a_{ij})$; and for key/value pair e_{ij}/r_{ij}, it issues an intermediate key/value pair e_{*j}/r_{ij}.

In the reduce stage, each process node collects all values associated with an intermediate key e_{i*} to generate new r_{i*} according to Equation (3.5), and all intermediate values associated with the same key e_{*j} to generate new a_{*j} according to Equations (3.6) and (3.7). Thus, the one time map-reduce process corresponds to one iteration in our affinity propagation algorithm.

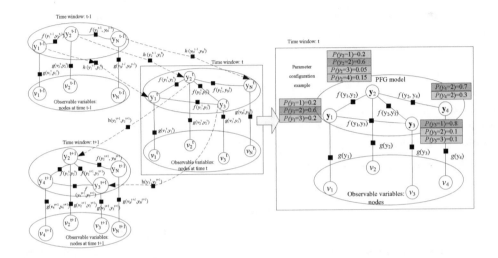

Figure 3.3: Graphical representation of the dynamic factor graph.

3.2.2 DYNAMIC SOCIAL INFLUENCE ANALYSIS

We further extend the social influence analysis to the dynamic setting, where the network structure and the users' topic distribution may change over time. In general, the input of dynamic social influence analysis is a serial of time-dependent social networks $\{G^t\} = \{(V^t, E^t)\}$, where V^t is a set of nodes (users, entities) appearing within the time window t and E^t is the set of directed/undirected edges. Each edge $e_{ij}^t \in E^t$ is associated with a weight/similarity w_{ij}^t, which can be defined in different ways, depending on the specific application.

We still use the factor graph model to learn the influence strength between users. The technical issue we need to address is how to incorporate the time information. Given T time-specific social networks $\{G^t\} = \{(V^t, E^t)\}_{t=1}^T$, DFG models the networks as a sequence of time-dependent factor graphs. At each time window, the factor graph has a similar structure with the PFG model. In addition, each factor graph also depends on the factor graph of the previous time window. Thus, the sequence of time-dependent factor graphs forms a Markov chain. Factor functions are defined between the variables of two consecutive time-dependent factor graphs.

A forward-backward message passing process is defined to capture the dependencies between networks of two time windows. Figure 3.3 shows the graphical representation of the DFG model.

To formally define the DFG model, we add a superscript t to the variables (v_i, y_i, Λ_{ijk}, and μ_{ij}) in PFG. Within the factor graph of each time window, the factor functions (node factor function and edge factor function) are defined similarly to the PFG model. Between two consecutive time-dependent factor graphs, we define a bridge factor function between y_i^t and y_i^{t+1}. In this way, the influence μ_{ji}^t (or $P(y_i^t = j)$) of user v_j on v_i not only is determined by their local and network structure information at time t, but also depends on their historic influence. Specifically, the bridge factor function as:

$$h(y_j^t, y_j^{t+1}) = \begin{cases} \frac{q}{|SC(j)|} & y_j^t = y_j^{t+1} \\ \frac{1-q}{|SC(j)|(|SC(j)|-1)} & y_j^t \neq y_j^{t+1} \end{cases}, \tag{3.11}$$

where $q \in [0, 1]$ is a weight, indicating the probability of one user's influence on another preserves with time changing. The weight q captures the time dependencies for the dynamic social influence analysis. It can be learned in a similar way as Algorithm 2.

Now the joint probability becomes

$$P(y_1^t, \ldots, y_N^t, v_1, \ldots, v_N) =$$
$$\prod_{v_i^t \in V^t} g(y_i^t, v_i^t) \prod_{e_{ij} \in E^t} f(y_i^t, y_j^t) \prod_{v_i^t \in V} h(y_i^{t-1}, y_i^t). \tag{3.12}$$

We generalize TAP learning algorithm to solve the learning problem for the DFG model. By introducing two new variables α and β, respectively, representing the forward and backward messages passed between two time-windows, we can obtain the following update rules:

$$\Lambda_{ijk}^t = \frac{1}{Z_i^t} \sum_{l \in SC(i)} f_{ij}(l,k) b_{il}^t \prod_{s \in NB(i)\setminus\{j\}} \Lambda_{sil}^t \tag{3.13}$$
$$\sum_{k' \in SC(i)} h(k',k)\alpha_{k'i}^{t-1} \sum_{k'' \in SC(i)} h(k,''k)\beta_{k''i}^{t+1}$$

$$\alpha_{ki}^t = \frac{1}{Z_i^t} \sum_{l \in SC(i)} h(l,k)\alpha_{li}^{t-1} b_{il}^t \prod_{s \in NB(i)} \Lambda_{sil}^t \tag{3.14}$$

$$\beta_{ki}^t = \frac{1}{Z_i^t} \sum_{l \in SC(i)} h(l,k)\beta_{li}^{t+1} b_{il}^t \prod_{s \in NB(i)} \Lambda_{sil}^t \tag{3.15}$$

$$Z_j^t = \sum_{l \in SC(j)} \alpha_{lj}^{t-1} \beta_{lj}^{t+1} b_{jl} \prod_{i \in NB(j)} \Lambda_{ijl}^t. \tag{3.16}$$

Thus, the probabilistic influence can be calculated by

$$\mu_{kj}^t = \frac{1}{Z_j^t} \alpha_{kj}^{t-1} \beta_{kj}^{t+1} b_{jk} \prod_{i \in NB(j)} \Lambda_{ijk}^t, \tag{3.17}$$

where α_{kj}^{t-1} is the message from the previous time window and β_{kj}^{t+1} is the message from the next time window. Above is the solution when influence scores in all time windows are calculated together. Given that the influence is evolving forward with time, which means $\{\mu_{ij}^t\}$ should be generated without information about $\{\mu_{ij}^{t+1}\}$, we can constraint the message passing process by only allowing forward message α and discarding β. Then it turns out we can compute the influence time by time, with a forward version of TAP.

Both the forward-backward and forward version of the TAP algorithm inherit the nice property of "local" update, which makes the algorithm easy to be parallelized. We implement the distributed learning under the map-reduce platform. Details are omitted due to space limitation.

3.2.3   MODEL APPLICATION

The social influence graphs by TAP can help with many applications. Here we illustrate two applications: expert finding and influence maximization.

Expert finding. Here we present three methods for expert finding: (1) PageRank+Language Model (PR); (2) PageRank with global Influence (PRI); and (3) PageRank with topic-based influence (TPRI).

Baseline: PR. One baseline method is to combine the language model and PageRank [141]. Language model is to estimate the relevance of a candidate with the query and PageRank is to estimate the authority of the candidate. There are different combination methods. The simplest combination method is to multiply or sum the PageRank ranking score and the language model relevance score.

Proposed 1: PRI. In PRI, we replace the transition probability in PageRank with the influence score. Thus we have

$$r[v] = \beta \frac{1}{|V|} + (1 - \beta) \sum_{v':v' \to v} r[v']p(v|v'). \tag{3.18}$$

In traditional PageRank algorithm, $p(v|v')$ is simply the value of one divides the number of outlinks of node v'. Here, we consider the influence score. Specifically, we define

$$p(v|v') = \frac{\sum_z \mu_{v'v}^z}{\sum_{v_j:v' \to v_j} \sum_z \mu_{v'v_j}^z}.$$

Proposed 2: TPRI. In the second extension, we introduce, for each node v, a vector of ranking scores $r[v, z]$, each of which is specific to topic z. Random walk is performed along with the co-author relationship between authors within the same topic. Thus, the topic-based ranking score is defined as:

$$r[v, z] = \beta \frac{1}{|V|} p(z_k|v) + (1 - \beta) \sum_{v':v' \to v} r[v', z]p(v|v', z), \tag{3.19}$$

where $p(z|v)$ is the probability of topic z generated by node v and it is obtained from the topic model; $p(v|v', z)$ represents the probability of node v' influencing node v on topic z; we define it as

$$p(v|v', z) = \frac{\mu^z_{v'v}}{\sum_{v_j : v' \to v_j} \mu^z_{v'v_j}}. \tag{3.20}$$

Influence maximization. Given a social network with the learned social influence on each social relationship, i.e., $G' = (V, E, \Omega)$, the objective is to find a small subset of nodes (seed nodes) from the network that could maximize the spread of influence. The problem of influence maximization has been proven to be a NP-hard problem [79]. A greedy approximation algorithm can guarantees that the influence spread is no worse than $(1 - 1/e)$ of the optimal influence spread. One major problem of the greedy algorithm is its low efficiency. Chen et al. [26, 27] developed new heuristics to accelerate the greedy algorithm. The social influence learned by the proposed TAP algorithm can provide the input to the greedy algorithm.

3.2.4   EXPERIMENTAL RESULTS

In this section, we present various experiments to evaluate the efficiency and effectiveness of the proposed approach. All data sets, codes, and tools to visualize the generated influence graphs are publicly available at `http://arnetminer.org/lab-datasets/soinf/`.

Experimental setup. We perform our experiments on three real-world data sets: two homogeneous networks and one heterogeneous network. The homogeneous networks are academic co-author network (Coauthor, for short) and paper citation network (Citation, for short). Both are extracted from academic search system Arnetminer.[1] The co-author data set consists of 640,134 authors and 1,554,643 co-author relations, while the citation data set contains 2,329,760 papers and 12,710,347 citations between these papers. Topic distributions of authors and papers are discovered using a statistical topic modeling approach, Author-Conference-Topic (ACT) model [148]. The ACT approach automatically extracts 200 topics and assigns an author-specific topic distribution to each author and a paper-specific topic distribution to each paper.

The other heterogeneous network is a film-director-actor-writer network (shortly Film), which is crawled from Wikipedia under the category of "English-language films."[2] In total, there are 18,518 films, 7,211 directors, 10,128 actors, and 9,784 writers. There are 142,426 relationships between the heterogeneous nodes in the dataset. The relationship types include: film-director, film-actor, film-writer, and other relationships between actors, directors, and writers. The first three types of relationships are extracted from the "infobox" on the films' Wiki pages. All the other types of people relationships are created as follows: if one people (including actors, directors, and writers) appears on another people's page, then a directed relationship is created between

[1]http://arnetminer.org
[2]http://en.wikipedia.org/wiki/Category:English-language_films

them. Topic distributions of the heterogeneous network is initialized using the category information defined on the Wikipedia page. More specifically, we take ten categories with the highest occurring times as the topics. The ten categories are: "American film actors," "American television actors," "Black and white films," "Drama films," "Comedy films," "British films," "American film directors," "Independent films," "American screenwriters," and "American stage actors." As for the topic distribution of each node in the film network, we first calculate how likely a node v_i belong to a category (topic) z, i.e., $p(v_i|z)$, according to $\frac{1}{|V_z|}$, where $|V_z|$ is the number of nodes in the category (topic) z. Thus, for each node, we will obtain a set $\{p(v_i|z)\}_{z=1}^T$ of likelihood for each node. Then we calculate the topic distribution $\{p(z|v_i)\}_{z=1}^T$ according to the Bayesian rule $p(z|v_i) \propto p(z)p(v_i|z)$, where $p(z)$ is the probability of the category (topic).

Evaluation measures. For quantitatively evaluate our method, we consider two performance metrics:

- **CPU time.** It is the execution elapsed time of the computation. This determines how efficient our method is.

- **Application improvement.** We apply the identified topic-based social influence to help expert finding, an important application in social network. This will demonstrate how the quantitative measurement of the social influence can benefit the other social networking application.

The basic learning algorithm is implemented using MATLAB 2007b and all experiments with it are performed on a Server running Windows 2003 with two Dual-Core Intel Xeon processors (3.0 GHz) and 8 GB memory. The distributed learning algorithm is implemented under the Map-Reduce programming model using the Hadoop platform.[3] We perform the distributed train on 6 computer nodes (24 CPU cores) with AMD processors (2.3 GHz) and 48 GB memory in total. We set the maximum number of iterations as 100 and the threshold for the change of r and a to $1e - 3$. The algorithm can quickly converge after 7–10 iterations in most of the times. In all experiments, for generating each of the topic-based social influence graphs, we only keep 1,000 nodes that have the highest probabilities $p(v|z)$.

Scalability performance. We evaluate the efficiency of our approach on the three data sets. We also compare our approach with the sum-product algorithm.

Table 3.2 lists the CPU time required on the three data sets with the following observations.

Sum-Product vs. TAP. The new TAP approach is much faster than the traditional sum-product algorithm, which even cannot complete on the citation data set.

Basic vs. Distributed TAP. The distributed TAP can typically achieve a significant reduction of the CPU time on the large-scale network. For example, on the citation data set, we obtain a speedup 15X. While on a moderate scaled network (the coauthor data set), the speedup of the distributed TAP is limited, only 3.6. On a relative smaller network (the film data set),

---

[3]http://hadoop.apache.org/

the distributed learning underperforms the basic TAP learning algorithm, which is due to the communication overhead of the Map-Reduce framework.

Distributed Scalability. We further conduct a scalability experiment with our distributed TAP. We evaluate the speedup of the distributed learning algorithm on the six computer nodes using the citation data set with different sizes. It can be seen from Figure 3.4(a) that when the size of the data set increase to nearly one million edges, the distributed learning starts to show a good parallel efficiency (speedupX3). This confirms that distributed TAP like many distributed learning algorithms is good on large-scale data sets.

Table 3.2: Scalability performance of different methods on real data sets. >10 hr means that the algorithm did not terminate when the algorithm runs more than 10 hours

Methods	Citation	Coauthor	Film
Sum-Product	N/A	> 10hr	1.8 hr
Basic TAP Learning	> 10hr	369s	57s
Distributed TAP Learning	**39.33m**	**104s**	148s

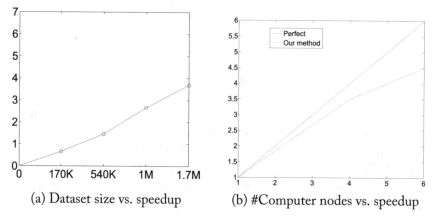

(a) Dataset size vs. speedup   (b) #Computer nodes vs. speedup

Figure 3.4: Speedup results.

Using our large citation data set, we also perform speedup experiments on a Hadoop platform using 1, 2, 4, 6 computer nodes (since we did not have access to a large number of computer nodes). The speedup, shown in Figure 3.4(b), show reasonable parallel efficiency, with a > 4× speedup using 6 computer nodes.

Quantitative case study. Now we conduct quantitatively evaluation of the effectiveness of the topic-based social influence analysis through case study. Recall the goal of expert finding is to identify persons with some expertise or experience on a specific topic (query) q. We define the baseline method as the combination [141] of the language model $P(q|v)$ and PageRank $r[v]$.

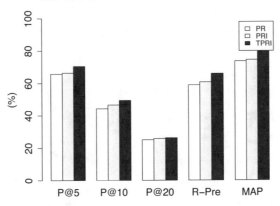

Figure 3.5: Performance of expert finding with different approaches.

We use an academic data set used in Tang et al. [141, 148] for the experiments. Specifically, the data set contains 14, 134 authors, 10, 716 papers, and 1, 434 conferences. Four-grade scores (3, 2, 1, and 0) are manually labeled to represent definite expertise, expertise, marginal expertise, and no expertise. Using this data, we create a coauthor network. The topic model for each author is still obtained using the statistical topic modeling approach [148]. With the topic models, we apply the proposed TAP approach to the coauthor network to identify the topic-based influences.

With the learned topic-based influence scores, we define two extensions to the PageRank method: PageRank with Influence (PRI) and PageRank with topic-based influence (TPRI). Details of the extension is described in Section 3.2.3. For expert finding, we can further combine the extended PageRank model with the relevance model, for example the language model by $P(q|v)r[v]$ or a topic-based relevance model by $\sum_z p(q|z)p(z|v)r[v, z]$, where $r[v]$ and $r[v, z]$ are obtained, respectively, from PRI and TPRI; $p(q|z)$, $p(z|v)$ can be obtained from the statistical topic model [141].

We evaluate the performance of different methods in terms of Precision@5 (P@5), P@10, P@20, R-precision (R-Pre), and mean average precision (MAP) [20, 33]. Figure 3.5 shows the result of expert finding with different approaches. We see that the topic-based social influences discovered by the TAP approach can indeed improve the accuracy of expert finding, which confirms the effectiveness of the proposed approach for topic-based social influence analysis.

Results of Dynamic Influences Analysis. The dynamic social influence analysis can benefit many applications. We use the influence maximization problem as two examples to demonstrate. The influence maximization problem is to find a small subset of nodes (seed nodes) in a social network that could maximize the spread of influence [42, 79, 121]. In most previous work, different algorithms were evaluated under simple assumptions about pairwise influence. Now the output of our social influence analysis can be used as the input of the influence maximization problem,

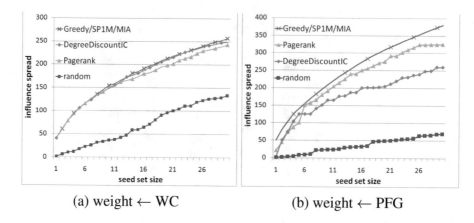

(a) weight ← WC　　　　　(b) weight ← PFG

Figure 3.6: Influence spread by different algorithms. PDF refers to the proposed dynamic influence analysis approach.

Table 3.3: Discovered seed nodes in influence maximization by the greedy algorithm with different influence schemes. *Unique*: influence=unique probability (0.01); *WC*: influence=inverse of in-degree; *PFG*: influence = result of our dynamic influence analysis

No.	Unique	WC	PFG
1	Philip S. Yu	Philip S. Yu	Jiawei Han
2	Jiawei Han	Jiawei Han	Qiang Yang
3	Christos Faloutsos	Wei Wang	Christos Faloutsos
4	Qiang Yang	Christos Faloutsos	Heikki Mannila
5	Heikki Mannila	Heikki Mannila	Vipin Kumar
6	Wei Wang	C. Lee Giles	C. Lee Giles
7	Jian Pei	Shusaku Tsumoto	Saso Dzeroski
8	Vipin Kumar	Jian Pei	Graham J. Williams
9	Bing Liu	Bing Liu	Myra Spiliopoulou
10	C. Lee Giles	Joost N. Kok	Eamonn J. Keogh
Overlap	0.4222	0.2444	0.1778

and we can test whether existing optimization algorithms perform as well as they do under the naive assumptions.

Figure 3.6 shows the solution found by several state-of-the-art algorithms defining the spread probability from v_i to v_j simply as $\frac{1}{d_j}$ (referred as WC model), where d_j is the in-degree of v_j. Beyond Greedy algorithm, we also test SP1M [80], using a simplified ICM model and MIA [26], a heuristic algorithm for general ICM model. Baseline algorithms include: (1) random, randomly picking seeds; (2) PageRank, selecting nodes with top PageRank score; and (3) De-greeDiscountIC, a heuristic algorithm with good performance in UICM [27]. Greedy algorithm provides best results as known. In WC model, SP1M and MIA perfectly match Greedy. De-greeDiscountIC has nearly matching performance. They all beat the other baselines. WC model presumes symmetric influence in common with UICM, while PFG does not. When PFG influ-ence score replaces the weights of WC, as shown in Figure 3.6, SP1M and MIA still approximate optimum, while DegreeDiscountIC degrades to Degree. Therefore, by applying our influence results, we find that SP1M and MIA are better approximation than DegreeDiscountIC if the symmetric assumption of influence does not hold.

Table 3.3 presents the discovered seed nodes by three different schemes to set the cascade influence scores. For each set of seed nodes, we calculate the *density* measure in network sci-ence, dividing the sum of coauthor papers by the number of different pairs between seeds, i.e., $\frac{10 \times 9}{2} = 45$ in our case. The larger is the overlap, the more concentrated are the seed nodes in the network. To maximize the influence spread, it is desirable to minimize the overlap. We see that our approach clearly outperforms the the other methods. It can be observed that Philip S. Yu does not appear as a top-10 seed in PFG model due to the large probability of him influenced by other seeds. In UICM and WC model, influence from neighbors to the node are independent, while in PFG, the correlation of influence between neighbors is captured.

3.2.5 SUMMARY

In this section, we introduce a novel problem of topic-based social influence analysis. We pro-pose a Topical Affinity Propagation (TAP) approach to describe the problem using a graphical probabilistic model. To deal with the efficient problem, we present a new algorithm for training the TFG model. A distributed learning algorithm has been implemented under the Map-reduce programming model. Experimental results on three different types of data sets demonstrate that the proposed approach can effectively discover the topic-based social influences. The distributed learning algorithm also has a good scalability performance. We apply the proposed approach to expert finding. Experiments show that the discovered topic-based influences by the proposed approach can improve the performance of expert finding.

3.3   MINING TOPIC-LEVEL INFLUENCE FROM HETEROGENEOUS NETWORKS

The proposed TAP method in Section 3 only considers homogeneous network with single type of objects, while many real-world networks usually consist of multiple heterogeneous objects. In addition, the TAP method assumes that a topic distribution is associated with each user, which is also not the case in real applications.

We propose the problem of topic-level influence mining from heterogeneous social net-works. The problem can be explained by using the example in Figure 3.7. The input (left figure) is a heterogeneous network consisting of web documents, users, and links between them. To leverage both content information of web documents and social network structure, we propose a proba-

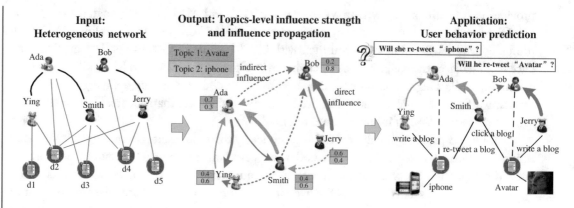

Figure 3.7: Problem illustration of mining topic-level influence in heterogeneous networks and predicting user behaviors.

bilistic generative model to jointly learn topics and to associate a topic distribution with each user which indicates his/her interests. Based on the modeling results, we can estimate the influence strength between friends. We further investigate two kinds of diffusion models for conservative and non-conservative influence propagations in social networks, which uncover the indirect influence between non-connected users. The middle figure of Figure 3.7 illustrates the output of topic discovery and influence propagation. The solid arrow indicates direct influence and the dashed arrow indicates indirect influence. The right figure illustrates a potential application, user behavior prediction, based on the learned influence.

In this section, we present a generative probabilistic model to quantify influence between users in heterogeneous social networks by utilizing both content and link information to mine direct influence strength in heterogeneous networks. We study two kinds of diffusion models for conservative and non-conservative influence propagations to learn indirect influence in social networks. We apply the discovered influence strength to user behavior prediction and validate how it can help some social applications. We conduct extensive experiments in four different types of data sets: Twitter,[4] Digg,[5] Renren,[6] and Cora,[7] and test the model performance in both qualitative and quantitative ways.

Problem Formulation.   Let us redefine the input of our problem.

Definition 3.2   [Heterogeneous Social Network] Define a network as $G = (V, D, E)$, where V is a set of user nodes, D is a set of document nodes, and E denotes a set of edges that includes social relationships connecting users and links connecting users and documents. For each edge

[4]http://www.twitter.com, a microblogging system.
[5]http://www.digg.com, a social news sharing and voting website.
[6]http://www.renren.com, one of the largest Facebook-like social networks in China.
[7]http://www.cs.umass.edu/mccallum/code-data.html, a bibliographic citation network

$e_{uv} = (u, v) \in E$, if there exists an edge between u and v, $e_{uv} = 1$; otherwise, $e_{uv} = 0$. The edges can be directed or undirected.

Many online social networks are heterogeneous consisting of different types of object nodes. For example, Twitter is comprised of users and microblogs. Digg consists of users and website URL addresses. Citation network consists of authors and publication papers. Here, we use "document" to represent different types of associated content (e.g., microblog, website, and paper) to each user. Thus, links in heterogeneous networks would contain friendships between users and authoring relationships between users and documents (links between documents are not considered in this chapter). The links can be directed or undirected. For example, in Twitter and citation networks, the links between users are directed from normal users to their followers. In Digg social network, the links between users are undirected. Furthermore, we assume that influence can propagate along social links, thus we have the following definition.

Definition 3.3   [Direct and Indirect Influence] Given two user nodes u, v in a heterogeneous network G, we denote $\delta_v(u) \in \{R^+ \cup 0\}$ as the influential strength of user u on user v. Furthermore, if $e_{uv} = 1$, we call $\delta_v(u)$ the direct influence of user u on v; if $e_{uv} = 0$, we call $\delta_v(u)$ the indirect influence of user u on v.

Direct influence indicates the influence between two users which are connected while indirect influence indicates the influence of two users which are not connected. Please note that influence is asymmetric, i.e., $\delta_v(u) \neq \delta_u(v)$. Based on the influence between node pairs, we can further define the concept of global influence.

Definition 3.4   [Global Influence] Given a heterogeneous network, $\Lambda(v) \in \{R^+ \cup 0\}$ is defined as the global influence of v, which represents the global influential strength of user v in the network.

The global influence strength has a close relationship with the (local) direct/indirect influence. For example, if a user has a strong influence on her/his friends, it is very likely that she/he has a strong global influence.

3.3.1   THE APPROACH FRAMEWORK

To summarize, we have two important intuitions for learning influence from heterogeneous social networks: (1) influence between users varies over different topics; and (2) user behaviors are not only influenced by their friends but also their n-degree friends (e.g., friends' friends). Indeed, in real networks users may be interested in different topics, e.g., in the research network an author may be interested in topics "database" and "data mining." The influential strength from one's coauthors on her/him w.r.t. the two topics might be very different. Actually, this has been qualitatively verified in sociology [56, 83] and quantitatively studied in Tang et al. [143]. More precisely, we can give the following descriptions for the intuitions.

1. Each node v is associated with a vector $\psi_v \in R^T$ of T-dimensional topic distribution ($\sum_z \psi_v(z) = 1$), where $\psi_v(z)$ indicates the interest probability of the node (user) on topic z.

2. Influence can propagate over social networks, thus the influence $\delta_v(u)$ of user u on v can be direct ($e_{uv} = 1$) or indirect ($e_{uv} = 0$).

3. The behavior of a user is either influenced by his/her friends who have the same behavior or generated depending on his/her interests.

The last intuition can be better explained by an example on Digg. A user may dig a story because his friends have digged this story or simply because he is interested in this topic.

From the technique perspective, our objective is to design a method to learn user interests (the associated topic distribution) and to estimate user influence simultaneously. In this book, we propose a topic-level influence modeling framework. First, by combining both textual information and link information in heterogeneous networks, we present a probabilistic generative model to learn user interests which are represented as mixtures of topics and direct influence between users simultaneously. Second, based on direct influence, we study two types of influence propagation mechanism, which are conservative and non-conservative influence propagations, to derive indirect influence between users.

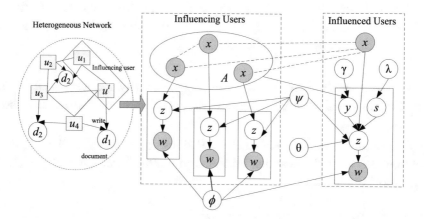

Figure 3.8: Probabilistic generative model to estimate direct influence strength.

Influence is interacted with many potential factors, e.g., similarity and correlation [2, 31]. Here we have two general assumptions in order to model the influence strength quantitatively.

Assumption 1 *Users with similar interests have a stronger influence on each other.*

This assumption actually corresponds to the influence and selection theory [2]. In real networks, the similarity can be calculated based on the content information associated with each user. Thus, influence can be represented as to which extent the textual content is "copied" from the influencing nodes. For example, in the citation network, if the content of document d_1 is very similar to that of document d_2, we may deem that d_1 "copies" a lot of ideas from d_2, thus d_1 is influenced by d_2 a lot.

Assumption 2 *Users whose actions frequently correlate have a stronger influence on each other.*

The co-occurrence frequency is often used to indicate the correlation strength between two nodes, which is denoted by the weights of edges in networks. Thus, the influence strength between two nodes would be enlarged by their frequent co-occurrence. For example, if author a cites a number of papers of author b, then a should be strongly influenced by b. For another example on Twitter, if user a replies or re-tweets many microblogs posted by user b, then it is very likely that b has a strong influence on a.

Based on these considerations, we propose a probabilistic generative model to jointly learn user interests and direct influence strength between users quantitatively.

Table 3.4: Variable descriptions

Notation	Description
x, x'	the influenced/influencing user
w, w'	words in the associated document
z, z'	topic assignment to each word
d, d'	document associated with influenced/influencing user
A_x	the user list who may influence x
y	the influencing user from A_x
s	the label denoting either influencing or not
W	the number of words in the data set
T	the number of topics to be extracted
θ	the topic mixture of influencing users
ψ	innovative topic mixture of users
ϕ	word distribution for each topic
γ	the influence mixture of users
λ	the parameter to draw the label s
α	the Dirichlet prior for hidden variables

A Probabilistic Generative Model. We propose a probabilistic model to mine topics and influence strength simultaneously. The model combines the content information and network structure in heterogeneous networks as shown in Figure 3.8. We assume that the behavior of each influenced user can be generated in two ways, either depending on his/her own interests or influenced by one of his/her friends. For example, when a user shares a blog on Renren, he/she may like its

foreach *influencing user x'* **do**
    foreach *associated document d'* **do**
        foreach *word $i \in d'$* **do**
            Draw a topic $z'_{d',i} \sim multi(\psi_x)$ from the topic mixture of user $x'_{d',i}$;
            Draw a word $w'_{d',i} \sim multi(\phi_{z_{d,i}})$ from $z'_{d',i}$-specific word distribution;
        end
    end
end
foreach *influenced user x* **do**
    foreach *associated documents d* **do**
        foreach *word $i \in d$* **do**
            Toss a coin $s_{d,i} \sim bernoulli(\lambda_{x_{d,i}})$, where
            $\lambda_{x_{d,i}} = p(s = 0|x_{d,i}) \sim beta(\alpha_{\lambda_{s_0}}, \alpha_{\lambda_{s_1}})$ which indicates the proportion between
            the innovation and influenced probability of $x_{d,i}$;
            if $s_{d,i} = 0$ **then**
                Draw a influencing user $y_{d,i} \sim multi(\gamma_x)$ from the user list A_x;
                Draw a topic $z_{d,i} \sim multi(\theta_y)$ from the topic mixture of $y_{d,i}$;
            end
            if $s_{d,i} = 1$ **then**
                Draw a topic $z_{d,i} \sim multi(\psi_x)$ from the topic mixture of $x_{d,i}$;
            end
            Draw a word $w_{d,i} \sim multi(\phi_{z_{d,i}})$ from $z_{d,i}$-specific word distribution;
        end
    end
end

Algorithm 5: Probabilistic generative process.

content or follow the action of one of his/her friends who also share it. Thus the proposed model consists of the following two parts, and the whole generative process are illustrated in Algorithm 5 (Table 3.4 lists the descriptions of variables).

- **User interest modeling** As shown in the middle part of Figure 3.8, each user x is represented as a multinomial distribution over topics ψ, which indicates user interests. We assume that topics of documents are generated based on user interests. Then each word w in documents is generated from one topic z selected from the distribution. The details of the generative process are illustrated in the first iteration of Algorithm 5.

- **Influence strength mining** The right part of Figure 3.8 illustrates influence strength modeling. The parameter s, which is generated from a Bernoulli distribution with parameter λ, is used to control the influence situation. We assume that when $s = 1$, the behavior is generated based on his/her own interests, while when $s = 0$, the behavior of the user is influenced by one of his/her friends. Then another parameter γ is used to indicate the influence strength from candidate user set A_x to user x, based on which one influencing user

y is selected from A_x. At last, a topic is generated from the mixture of topics of a user—the user himself/herself x or one of his/her friends y, based on which the word w is generated. This part corresponds to the second iteration of Algorithm 5.

In the above generative process, A_x is the candidate influencing user set w.r.t. x, thus A_x changes with x. Besides, A_x is determined by real applications, which considers both directed and undirected links between users. For example, in Twitter network A_x denotes the users whom a blog is re-tweeted from while in citation networks it denotes the authors of cited papers. In these networks, the links between users are directed. In some other networks, such as Renren and Digg, A_x denotes the friends of user x who also share or dig the same story, and the links are undirected. Thus, the proposed model is able to handle both types of cases.

Model Learning via Gibbs Sampling. We employ Gibbs sampling to estimate the model. Gibbs sampling is an algorithm to approximate the joint distribution of multiple variables by drawing a sequence of samples, which iteratively updates each latent variable under the condition of fixing remaining variables. We list the update equations for each variable as below and the details of derivation can refer to the appendix. In all the update equations, $N(*)$ is the function which stores the number of samples during Gibbs sampling. For example, $N_{x,z,s}(x, z, 1)$ represents the number of samples of topics z which are supposed to be generated from user x when $s = 1$:

$$p(s_i = 0|\vec{s}_{-i}, x_i, z_i, .) \propto$$
$$\frac{N_{x',z'}(y_i,z_i)+N_{y,z,s}(y_i,z_i,0)+\alpha_\theta}{N_{x'}(y_i)+N_{y,s}(y_i,0)+T\cdot\alpha_\theta} \cdot \frac{N_{x,s}(x_i,0)+\alpha_{\lambda_{s_0}}}{N_x(x_i)+\alpha_{\lambda_{s_0}}+\alpha_{\lambda_{s_1}}} \tag{3.21}$$

$$p(s_i = 1|\vec{s}_{-i}, x_i, z_i, .) \propto$$
$$\frac{N_{x,z,s}(x_i,z_i,1)+\alpha_\psi}{N_{x,s}(x_i,1)+T\cdot\alpha_\psi} \cdot \frac{N_{x,s}(x_i,1)+\alpha_{\lambda_{s_1}}}{N_x(x_i)+\alpha_{\lambda_{s_0}}+\alpha_{\lambda_{s_1}}} \tag{3.22}$$

$$p(y_i|\vec{y}_{-i}, s_i = 0, d_i, x_i, z_i, A_x, .) \propto$$
$$\frac{N_{x,y,s}(x_i,y_i,0)+\alpha_\gamma}{N_{x,s}(x_i,0)+|A_x|\cdot\alpha_\gamma} \cdot \frac{N_{x',z'}(y_i,z_i)+N_{y,z,s}(y_i,z_i,0)+\alpha_\theta}{N_{x'}(y_i)+N_{y,s}(y_i,0)+T\cdot\alpha_\theta} \tag{3.23}$$

$$p(z_i|\vec{z}_{-i}, s_i = 0, w_i, .) \propto$$
$$\frac{N_{x',z'}(y_i,z_i)+N_{y,z,s}(y_i,z_i,0)+\alpha_\theta}{N_{x'}(y_i)+N_{y,s}(y_i,0)+T\cdot\alpha_\theta} \cdot \frac{N_{w,z}(w_i,z_i)+N_{w',z'}(w'_i,z'_i)+\alpha_\phi}{N_z(z_i)+N_{z'}(z_i)+W\cdot\alpha_\phi} \tag{3.24}$$

$$p(z_i|\vec{z}_{-i}, s_i = 1, w_i, .) \propto$$
$$\frac{N_{x,z,s}(x_i,z_i,1)+\alpha_\psi}{N_{x,s}(x_i,1)+T\cdot\alpha_\psi} \cdot \frac{N_{w,z}(w_i,z_i)+N_{w',z'}(w'_i,z'_i)+\alpha_\phi}{N_z(z_i)+N_{z'}(z_i)+W\cdot\alpha_\phi}. \tag{3.25}$$

Through the Gibbs sampling process, we obtain the sampled coin s_i, influencing user y_i, and topic z_i for each word. Then the influence strength can be estimated by Equation (3.26), which are averaged over the sampling chain after convergence. K denotes the length of the sampling chain:

$$\delta_x(y) = \gamma_x(y) = \frac{1}{K}\sum_{i=1}^{K}\frac{N_{x,y,s}(x, y, 0)^i + \alpha_\gamma}{N_{x,s}(x, 0)^i + |A|\cdot\alpha_\gamma}. \tag{3.26}$$

The equation is consistent to our assumptions in a statistical way. Take citation networks for example. If author x cites more papers of author y and "copies" more content from y, $N_{x,y,s}(x, y, 0)$ will be larger, and thus the influence from y to x will be stronger. Besides, it is easy to get that $\sum_{y=1}^{|A_x|} \delta_x(y) = 1$, i.e., the sum of influence on user x from all the users obtained in the model equals to 1. And the model does not consider the influence between the nodes which are not connected, i.e., $\delta_x(y) = 0$ when x and y are not connected.

Furthermore, we can estimate the topic-level influence strength. Suppose $\delta_{x,z}(y)$ represents the influence strength from user y to user x on the topic z, which satisfy that $\delta_x(y) = \sum_{z=1}^{T} \delta_{x,z}(y)$. Thus, the topic-level influence can be estimated by Equation (3.27).

$$\delta_{x,z}(y) = \frac{1}{K} \sum_{i=1}^{K} \frac{N_{x,y,z,s}(x, y, z, 0)^i + \frac{1}{T} \cdot \alpha_\gamma}{N_{x,s}(x, 0)^i + |A| \cdot \alpha_\gamma}. \tag{3.27}$$

3.3.2   INFLUENCE PROPAGATION AND AGGREGATION

The above probabilistic model only discovers direct influence, but does not consider indirect influence. In reality, like information or virus, influence also propagates over networks, which produces different types of indirect influence. Take Figure 3.9(a) for example. If $a1$ influences $a2$ and $a2$ influences $a3$, then $a1$ will influence $a3$ potentially, i.e., two-degree of influence. Figure 3.9(b) demonstrates the influence enhancement: if $a1$ influences $a3$ and $a4$ while $a3$ and $a4$ also have an influence on $a2$, then the influence from $a1$ to $a2$ should be enhanced. We further study atomic and iterative influence propagation over social networks in this section, via which indirect influence can be obtained from direct influence and global influence strength can be estimated.

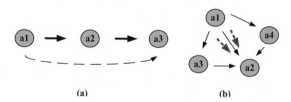

(a)   (b)

Figure 3.9: Influence propagation.

Atomic Influence Propagation. As shown in Figure 3.9, we observe there are two basic processes for influence propagation.

- **Concatenation** The indirect influence from $a1$ to $a3$ in Figure 3.9(a) can be modeled as a concatenate result of the direct influence from $a1$ to $a2$ and the direct influence from $a2$ to $a3$.

- **Aggregation** The enhancement of the influence from $a1$ to $a2$ in Figure 3.9(b) can be defined as an aggregate result of the direct influence among the neighborhood of $a1$ and $a2$.

Therefore, the *atomic influence propagation* is defined as:

$$\delta_v(u) = \Diamond(\forall w \in Nb(v) : \delta_v(w) \circ \delta_w(u)),\qquad(3.28)$$

where $Nb(v)$ is the set of neighbors of node v, $\circ$ is the concatenation function, and $\Diamond$ is the aggregation function.

In real processes, multiplication operation or minimum value is often used as concatenation function while addition operation or maximum value is used as the aggregation function. In particular, if we employ multiplication and addition operations to replace the concatenation and aggregation function in Equation (3.28), respectively, then the atomic influence propagation can be instantiated as:

$$\delta_v(u) = \sum_{w \in Nb(v)} \delta_v(w) \cdot \delta_w(u).\qquad(3.29)$$

Suppose Δ_v represents the vector of the influence strength from all the nodes in the network on node v, i.e., $\Delta_v = (\delta_v(u_1), \delta_v(u_2), ..., \delta_v(u_n))$. And we use superscript to denote the propagation step, i.e., Δ^0 denotes the initial influence strength and Δ^1 denotes the influence strength after the atomic propagation. Then the atomic influence propagation can be represented as the matrix multiplication.

$$\Delta_v^1 = \Delta_v^0 \cdot M\qquad(3.30)$$

where M is the transition matrix and $M = (\Delta_{v_1}; \Delta_{v_2}; \ldots; \Delta_{v_n})$, i.e., each element in the transition matrix $M(v, u) = \delta_v(u)$.

Iterative Influence Propagation. In reality, the indirect influence along longer paths, e.g., three-degree or four-degree influence, also have effect on the nodes in a network. In other words, influence can propagate iteratively to collect the contribute of influence on longer paths. Thus, the atomic influence propagation should be performed iteratively to propagate direct influence over the entire network. Thus, the influence strength on k-length paths can be calculated by k steps of atomic propagations.

If the atomic propagation is defined as Equation (3.30), the influence strength vector after k-step atomic propagation can be calculated by the matrix powering operation.

$$\Delta_v^k = \Delta_v^{k-1} \cdot M = \Delta_v^0 \cdot M^k,\qquad(3.31)$$

where $M^k = M^{k-1} \cdot M$. Δ^k denotes the influence strength vector on k-length paths.

Formally, we define the *iterative influence propagation* as follows:

- enumerate all paths between each two nodes;

- calculate the influence propagation strength on each path via a concatenation function; and

- combine the influence strength on all the paths via an aggregation function.

Suppose the final influence strength between two nodes after k-step iterative propagation is denoted as Δ^{f_k}. Based on the above definition, it should collect all the contributes of the influence strength on paths with the length ranging from 0 to k, i.e.,

$$\Delta^{f_k} = \Diamond(\forall i \in \{0, 1, 2, \ldots, k\} : \Delta^i). \tag{3.32}$$

If addition operation is used as the aggregation function, Δ^{f_k} can be inferred from the sequences of propagation via a weighted linear combination [60]:

$$\Delta^{f_k} = \sum_{i=0}^{k} \beta_i \Delta^i, \tag{3.33}$$

β_i denotes the weight for the influence strength on i-length paths, i.e., Δ^i.

Intuitively, the effect of the influence on shorter paths should be larger than the one on longer paths as the iterative propagation process brings in more outside information. Technically, β_i should decrease with the increase of iteration step i. Different strategies can be employed to assign the weights. In the next section, we will study two kinds of strategies, which are conservative propagation and non-conservative propagation, respectively.

Global Influence Estimation. Global influence is to measure one's influential ability over the whole network. For example, some authors are very influential on the topic of "data mining." In this section, we propose one way to estimate one node's global influence over the whole network.

Intuitively, the global influence of one node $\Lambda(u)$ should be related to its influence on all the other nodes in the network. If one node strongly influences many other nodes, its global influence might be also strong. Therefore, the global influence of a node is defined as an aggregation of its influence on the other nodes, specifically,

$$\Lambda(u) = \sum_{v} \delta_v(u). \tag{3.34}$$

The influence scores $\delta_v(u)$ include both direct and indirect influences.

3.3.3   CONSERVATIVE AND NON-CONSERVATIVE PROPAGATION

In this section, we describe two types of diffusion process—conservative and non-conservative diffusion processes—based on which we propose two kinds of methods to propagate influence over the network and to obtain indirect influence strength.

First, we formally define a propagation process over a network.

Definition 3.5 **[Propagation Process]** A propagation process over a network G is defined as a function $\{F_t(w) : (R^+ \cup \{0\})^{|V|} \rightarrow (R^+ \cup \{0\})^{|V|}\}$, where V is the set of nodes in G. w is a V-dimensional vector, which represents a weight distribution over the nodes in the network. t denotes propagation step.

Therefore, in a propagation process, each node in a network is first initialized with some mass, which is denoted as the weight of the node. Then via each step of propagation, some nodes transfer a part of the weights to their neighbors. Thus, through a t-step propagation process, a $|V|$-dimensional non-negative vector is mapped to another $|V|$-dimensional non-negative vector. In particular, when $t = 1$, the propagation is atomic propagation.

Conservative Propagation

Definition 3.6 **[Conservative Propagation]** For a propagation process F, if $\forall w \in (R^+ \cup \{0\})^{|V|}$, $||w||_1 = ||F(w)||_1$, i.e., it preserves the sum of the entries, we call the propagation process conservative propagation.

Therefore, conservative propagation simply redistributes the weights among the nodes in the network and keeps the sum of weights constant. There are many conservative propagation examples in the real world. Take the circulation of money for example. At each step of propagation, some nodes transfer a fraction of their money to their neighbors. But the total money in the network does not change. Traffic transportation and energy cycle are also conservative propagations as the total traffic or energy does not change with the propagation process.

Mathematically, random walk is a canonical example of conservative propagation. In a random walk, a particle starts to locate on a node. Then at each step, the particle selects one of the out-neighbors at random and moves to that node. A weight vector is used to represent the probability with which the particle can be found on each node. Thus, the sum of the weights equals to one. And after iterative propagations, the probabilities of finding the particle on the nodes change, but the sum remains to be one all the time.

PageRank is a classical random walk model, which is represented as:

$$pr(w) = (1 - \beta) \cdot w_0 + \beta \cdot pr(w) \cdot M, \qquad (3.35)$$

M is a transition matrix, in which the element $M(a, b)$ denotes the transfer probability from node a to b. β is a damping factor which is used to ensure the stationary probability distribution of the propagation. $1 - \beta$ is the restart probability, which gives the probability distribution when the random walk transition restarts. w_0 is the initial weight distribution, which is usually set to be uniform vector. Personalized PageRank [71] extends the model by setting w_0 to be a non-uniform starting vector.

Conservative Influence Propagation. We model the *conservative influence propagation* as a personalized PageRank in a network as Equation (3.36):

$$\Delta^{f_t} = (1 - \beta) \cdot \Delta^0 + \beta \cdot \Delta^{f_{t-1}} \cdot M. \tag{3.36}$$

The propagation probability matrix M can be set in various ways. If we use direct influence strength to define the propagation probability, i.e., $M(v, u) = \delta_v^0(u)$, then $\sum_u M(v, u) = 1$. It is easy to prove that the sum of influence strength from all the nodes on one node v remains to be one after influence propagation, i.e., $||\Delta_v^{f_t}||_1 = 1$. Thus, Equation (3.36) defines a conservative influence propagation.

This conservative influence propagation provides a strategy for the combination process in the iterative propagation. From Equation (3.36), it is easy to get that

$$\Delta^{f_t} = (1 - \beta) \cdot \Delta^0 \cdot \sum_{i=0}^{t-1} (\beta^i \cdot M^i) + \Delta^0 \cdot \beta^t \cdot M^t. \tag{3.37}$$

As the influence vector on t-length path is $\Delta^t = \Delta^0 \cdot M^t$,

$$\Delta^{f_t} = (1 - \beta) \cdot \sum_{i=0}^{t-1} (\beta^i \cdot \Delta^i) + \beta^t \cdot \Delta^t. \tag{3.38}$$

Thus, the conservative influence propagation defined in Equation (3.36) assigns different weights to the influences on different-length paths.

β is a damping factor, i.e., $0 \le \beta \le 1$. Thus, when t increases, β^t decreases, which makes the effect of influence on longer paths smaller.

Non-conservative Propagation.

Definition 3.7   [Non-conservative Propagation] For a propagation process F, if $\exists w \in (R^+ \cup \{0\})^{|V|}, ||w||_1 \ne ||F(w)||_1$, we call the propagation process non-conservative propagation.

Compared with conservative propagation, non-conservative propagation does not keep the sum of weights constant. There are also many non-conservative propagation examples in the real world. Take the spread of a virus for example. Suppose a virus is propagating over the social network. When one infected node infects its neighbors, it is still infected. Thus, the total number of infected nodes is increased with time. Therefore, the spread of virus is a kind of non-conservative process. Besides, information diffusion and oral advertising are also non-conservative propagations as the number of nodes which accept the information or advertisement increases with propagation step.

Alpha-Centrality, which was introduced by Bonacich [16, 17], can be used to model non-conservative propagation. The Alpha-Centrality vector $c(w)$ is defined as the solution of the following equation:

$$c^t(w) = w_0 + \beta \cdot c^{t-1}(w) \cdot M, \tag{3.39}$$

β is a damping factor. The starting vector w_0 is usually set to be in-degree centrality and M uses the adjacency matrix.

When $\beta < \frac{1}{|\lambda_1|}$ (where λ_1 is the largest eigenvalue of M), we can get that $c(w) = w_0 \cdot (I - \beta M)^{-1}$, where I is the identity matrix of size n. Using the identity

$$\sum_{t=1}^{\infty}(\beta^t \cdot M^t) = (I - \beta \cdot M)^{-1} - I \tag{3.40}$$

we can get

$$c(w) = w_0 \cdot (I - \beta \cdot M)^{-1} = w_0 \cdot \sum_{t=0}^{\infty}(\beta^t \cdot M^t). \tag{3.41}$$

Besides Alpha-Centrality, Katz score [77], SenderRank [81], and eigenvector centrality [15] are other examples of non-conservative mathematical metrics.

We model the *non-conservative influence propagation* process in the form of Alpha-Centrality as Equation (3.42):

$$\Delta^{f_t} = \Delta^0 + \beta \cdot \Delta^{f_{t-1}} \cdot M. \tag{3.42}$$

For Alpha-Centrality, M is usually set to be adjacency matrix. Here we also use direct influence strength to define the transition matrix M, i.e., $M(v, u) = \delta_v(u)$. It is easy to prove that the sum of influence strength from all the nodes on node v increases with non-conservative propagation step, i.e., $||\Delta_v^{f_t}||_1 > 1$. Thus, Equation (3.42) defines a non-conservative propagation for local influence.

This non-conservative influence propagation provides another strategy for the combination process in the iterative propagation. From Equation (3.42), we can get

$$\Delta^{f_t} = \Delta^0 \cdot \sum_{i=0}^{t}(\beta^i \cdot M^i) = \sum_{i=0}^{t} \beta^i \cdot \Delta^i. \tag{3.43}$$

Thus, it assigns different weights to the influence strength on different-length paths. However, the weight assignment strategy is different from conservative propagation referring to Equation (3.38).

Both conservative and non-conservative influence propagations collect all the contributes of direct and indirect influence on the propagating paths. And both of them define a weight assignment strategy to distinguish the effect of influence on different-length paths. The major difference between these two types of models is that conservative propagation keeps the sum of influence in the whole network constant while non-conservative propagation does not.

Intuitively, indirect influence strength on shorter-paths should be more reliable since there have been fewer propagation steps. The more iteration steps, the more outside information will be brought. Thus, both conservative and non-conservative propagations utilize a damping factor β to penalize larger t-step propagations. As $0 \leq \beta \leq 1$, when t increases, β^t decreases greatly, which makes the effect of influence on $(t + 1)$-length paths very small. In another word, we do not need to iterate influence propagation for many times to obtain the final indirect influence, i.e., t can be set as a small number. Besides, when $\beta = 0$, both conservative and non-conservative propagations only utilize direct influence and ignore the effect of indirect influence.

3.3.4   USER BEHAVIOR PREDICTION

The learned influence strength can be used to help with many applications. Here we illustrate one application on user behavior prediction, i.e., how the learned influence can help improve the performance of user behavior prediction.

We evaluate our approach for user behavior prediction on Renren, Twitter, and Digg. The user behavior is defined as one time connection between a user and a document. We here take Digg as the example for explanation. Intuitively, if more friends of a user dig a story, there is a larger probability that the user will also dig it. Thus, a vote-based relational neighbor classifier [102] can be used as a baseline. Then, we use the influence strength obtained from our approach to distinguish different friends' weights and estimate the probability of users' digging stories as follows:

$$p(d|u) = \frac{1}{\sum_v \delta_u(v)} \sum_{v \in Nb(u)} \delta_u(v) p(d|v) \tag{3.44}$$

where $Nb(u)$ denotes the friends of u.

Besides, the similarity between users can also be used to distinguish different friends' weights in the above intuitive method for prediction. Thus the prediction probability is estimated as Equation (3.45) for comparison, where the similarity between users $s(v, u)$ is calculated as the Euclidean distance of user distributions over topics:

$$p(d|u) = \frac{1}{\sum_v s(v, u)} \sum_{v \in Nb(u)} s(v, u) p(d|v). \tag{3.45}$$

We will test the user behavior prediction performance based on the above three methods in the following experiments and demonstrate the effect of influence strength obtained from both conservative and non-conservative influence propagations for social network applications.

3.3.5   EVALUATION

In this section, we present various experiments to evaluate the efficiency and effectiveness of the proposed approach. The data sets and codes are publicly available.[8]

Experimental Setup

Data Sets. We prepare four different types of heterogeneous networks for our experiments, including Renren, Twitter, Digg, and Citation networks. Renren is a very popular FaceBook-style social website in China, on which users (especially the undergraduate and graduate students) connect with their classmates or friends and share interesting web content. Twitter is a microblog website, on which users can publish blogs and re-tweet friends' blogs. Digg is a different type of social website, on which users can submit, dig and comment on stories. Users also have links to their friends, which indicate their relationship. We collect user relationship and document content from these websites.

- **Renren social network.** The data contains 5,000 users and the web content shared by these users in one month which includes about 10,000 documents and 30,000 words.

- **Twitter social network.** The dataset includes about millions of microblogs related to about 40,000 users and 50,000 keywords (removing the stop words and the infrequent words).

- **Digg social network.** The data contains about 1 million stories related to 10,000 users and 30,000 keywords, in which we aim to mine user influence as well.

- **Citation network.** We crawled the citation data of about 1,000 documents from the Internet on several specific topics, e.g., "topic models," "sentiment analysis," "association rule mining," "privacy security," etc. Besides, the public citation data set Cora is also used in our experiments.

We apply our model to the above four data sets. The algorithms are implemented in C++ and run on an Intel Core 2 T7200 and a processor with 2 GB DDR2 RAM. The parameters of the model will be discussed in the following subsections.

Evaluation Aspects. We evaluate our method on the following three aspects.

- **Influence strength prediction:** As it is more intuitive and easier for people to distinguish the influence strength in citation networks, we manually label the citation data and then test the influence prediction performance in it. We compare the results of our approach with previous work [39] to demonstrate our model's better performance in terms of influence prediction.

- **User behavior prediction:** We use the derived influence strength to help predict user behaviors and compare the prediction performance with that of baseline as well as the method

[8]http://arnetminer.org/heterinf

based on user similarity, as described in Section 3.3.4. The results demonstrate how the quantitative measurement of the influence can benefit social network applications.

- **Topic-level influence case study:** We show several case studies to demonstrate concrete influence weights between users and show how effectively our method can identify topic-level influence. In particular, we study the global influence of authors in citation networks to demonstrate semantic meaning of topic-level influence. And we compare the results with that of previous work [143] which can also be used to mine topic-level influence to demonstrate the better performance of our approach.

Results

Influence Strength Prediction. In Dietz et al. [39], researchers evaluated the document influence prediction performance in a manually labeled data set. We use the same data from the authors and also test the influence prediction performance of our model in it. However, the data set, which only contains 22 citing documents and 132 documents in all, is so small that the results could be ad-hoc sometimes. Therefore, besides using this data, we also manually label document influence strength in a larger data set with about 1,000 documents. We classify the influence strength into three levels: 1, 2, 3. Similar to Dietz et al. [39], we use the quality measure, averaged AUC (Area Under the ROC Curve) values for the decision boundaries "1 vs. 2, 3" and "1, 2 vs. 3" for each citing document, to evaluate the prediction performance.

Figure 3.10 shows the comparative results in these two data sets, where Data1 is the small data set obtained from Dietz et al. [39] while Data2 is our larger labeled data set. $M1$ and $M2$ are used to denote our model and the model of Dietz et al. [39], respectively. And we use the real and dash lines to distinguish the results of these two models in the figure. We calculate all the AUC values with the number of topics changing from 10 to 50. Thus, this figure demonstrates that in the small data set our model can achieve as good prediction performance as the work of Dietz et al. [39] while in the larger data set, our prediction performance is better than theirs.

Furthermore, we compare the influence prediction performance before and after influence propagation in our labeled data set. The results prove that the influence prediction performance is enhanced after influence propagation (AUC values are enhanced from 0.69 to o.76). Moreover, the influence prediction performance is robust to the parameters t and β. In particular, when t changes, the performance changes little, which is consistent to the observation in Figure 2.8. It means that influence does propagate over the network, but the effect of propagation is reduced with propagation step.

User Behavior Prediction. We employ our model to discover the concrete influence strength between the 5,000 users in Renren social networks. Then we apply the learned influence strength to user behavior prediction, as described in Section 3.3.4. In particular, the parameters which are the damping factor β and iteration step t for both conservative and non-conservative influence propagations are varied to test the effect of influence propagation process. About 36,000 tuples

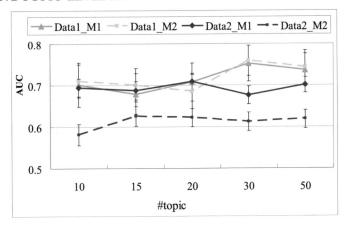

Figure 3.10: Influence prediction performance comparison.

Table 3.5: Conservative and non-conservative influence propagation effect on user behavior prediction

method p	baseline	D	steps	influence propagation					
				$\beta = 0.3$		$\beta = 0.5$		$\beta = 0.8$	
				CIP	NCIP	CIP	NCIP	CIP	NCIP
average	0.101	0.160	$t = 1$	0.168	0.168	0.168	0.168	0.172	0.168
			$t = 5$	0.168	0.168	0.170	0.170	0.180	0.175
			$t = 10$	0.168	0.168	0.170	0.170	0.180	0.178
variance	0.011	0.048	$t = 1$	0.044	0.045	0.041	0.044	0.039	0.042
			$t = 5$	0.044	0.045	0.042	0.043	0.041	0.041
			$t = 10$	0.044	0.045	0.043	0.042	0.041	0.041

in Renren data set are used as testing samples. Each tuple represents that a user shares a web document, whose probability is estimated as Equation (3.44).

The average and variance values of the predicted probabilities for all the samples are calculated and shown in Table 3.5, where DI denotes direct influence, CIP and NCIP denote conservative and non-conservative influence propagations, respectively. The results demonstrate that using influence, especially the propagated influence, can greatly improve the predicted probabilities. But the parameters t and β as well as the propagation mechanism do not affect the probabilities a lot.

Then given a threshold, we calculate the prediction precision, which means how many testing samples' probabilities are larger than the threshold. Figure 3.11 shows four curves of prediction precision changing with the threshold in Renren data set, which indicate the performance of baseline, using direct influence without influence propagation, conservative, and non-conservative influence propagations with parameter $\beta = 0.8, t = 5$, respectively. The results demonstrate that influence-based behavior prediction approach outperforms the baseline. Thus, it proves that the

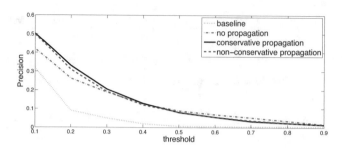

Figure 3.11: User behavior prediction precision on Renren network.

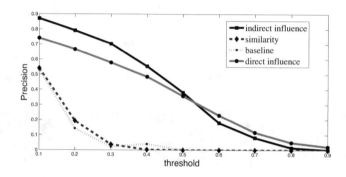

Figure 3.12: User behavior prediction precision on Digg network.

influence obtained from our model benefits the user behavior prediction greatly. Moreover, both conservative and non-conservative influence propagations improve the prediction precision and almost achieve the same performance.

Besides, we apply our model to the application of user behavior predication in Twitter and Digg social networks. In this experiment, we employ non-conservative influence propagation with $t = 5, \beta = 0.8$ to obtain indirect influence. We randomly select about 3,000 tuples from Digg and Twitter data sets as testing samples and estimate their probabilities. Table 3.6 shows the average and variance values of the predicted probabilities for all the samples. The prediction precision curves for these two data sets are shown in Figures 3.12 and 3.13, respectively. The results demonstrate that influence-based behavior prediction approach outperforms the baseline and the similarity-based method. In particular, it shows that influence propagation process enhances the user behavior prediction performance in Digg social network but it takes little effect in Twitter social network. Furthermore, comparing these two figures, we can get that the effect of influence in Digg social network is larger than that in Twitter social network. The conclusion is consistent to the observation in Figure 2.8.

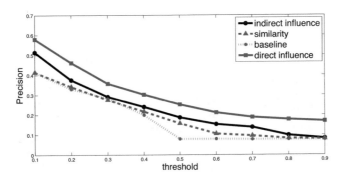

Figure 3.13: User behavior prediction precision on Twitter network.

Table 3.6: Behavior prediction probability

	Digg Social Network			
method$\diagdown$ p	baseline	similarity	DI	NCIF
average	0.112	0.121	0.366	0.405
variance	0.006	0.008	0.075	0.048
	Twitter Social Network			
method$\diagdown$ p	baseline	similarity	DI	NCIF
average	0.215	0.222	0.319	0.310
variance	0.078	0.089	0.129	0.136

Topic-level Influence Case Studies

Case Study 1: Topic-level influence graph. We apply our model to the citation network which we crawled from the Internet and set the number of topics to be 10 empirically. Figure 3.14 demonstrates the influence relationship between the papers on the topic "statistical topic models." The color bars show the topic distributions of these documents. In order to show the major influencing nodes clearly, we rank the influencing nodes according to each influenced node based on the influence strength and only display the top 2 most influencing ones in this figure. Thus, we can get that the top 2 most influencing documents on document "LDA" are "PLSA" and "variational inference." Furthermore, the results demonstrate that there are many documents which are most influenced by "LDA," e.g., "the author-topic model," "correlated topic model," "dynamic topic model," etc. Besides the influence from "LDA," strong influences also exist among these documents, e.g., "author-topic model" influences "author-recipient-model" strongly while "correlated topic model" influences "dynamic topic model" a lot.

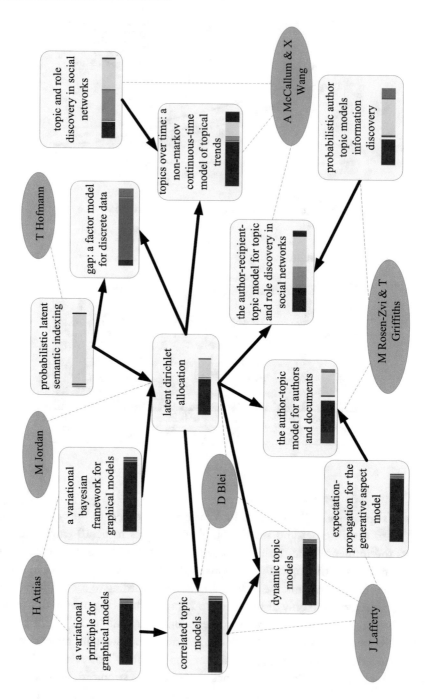

Figure 3.14: Document influence case study.

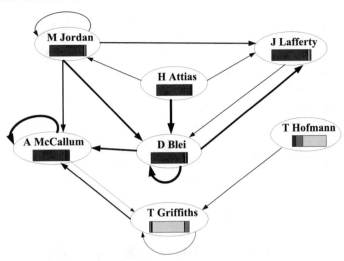

Figure 3.15: Author influence case study.

Figure 3.14 also shows the connections between authors and documents by dash lines. The influences between these authors are visualized in Figure 3.15. We only draw the lines when the pointing nodes are the top five most influencing authors on the pointed nodes. The thickness of the lines indicates the influence strength. From the results, we can get some meaningful conclusions. For example, Jordan is one of the most influential researchers to Blei. Although "PLSA" strongly influences "LDA" as Figure 3.14 shows, Hofmann does not have a great influence on Blei. The reason is that the area of Hofmann varies from the area of Blei (this can be observed from the topic distributions represented by colored bars) and furthermore Blei only cited few documents of Hofmann, i.e., correlation value is small. Other interesting results are also obtained, e.g., the influence of Blei on Lafferty is larger than the influence of Lafferty on Blei. Besides, the self-loop lines which indicate the self-influence show Jordan and Blei influence themselves greatly.

Table 3.7: Author ranking on "statistical topic models"

Direct Influence	Indirect Influence		Pagerank
	$t = 1$	$t = 5$	
TM Cover	D Blei	D Blei	M Jordan
A McCallum	A McCallum	A McCallum	D Blei
D Blei	TM Cover	M Jordan	J Lafferty
M Jordan	M Jordan	TM Cover	A McCallum
P Kantor	P Kantor	P Kantor	Z Ghahramani

Case Study 2: Topic-level global influence illustration. Table 3.7 shows an example of author ranking by estimated global influence on "statistical topic models" (t denotes the number of propagation steps). The results are very meaningful. If one node has a high reputation over the whole network, it can be treated as a key node which is very influential over the whole network. In another word, authority of one node can also be used to represent its global influence from some point of view. Therefore, we can employ PageRank [65, 116] over topic-level networks to estimate the nodes' global influence on one topic. The author ranking based on the authority from PageRank is also illustrated. We calculate the correlation coefficients between the global influence values estimated in the two ways, which ranges from 0.8–0.9 when the number of topics and iteration change. It proves that estimating global influence based on our framework can get highly-correlated results with PageRank authority. Thus, to some extent, it demonstrates that the influence discovered by our model is consistent to the global characteristics of the whole network structure.

Table 3.8: Influence aggregation values on topics

Topic	OODB	IR	DM	DBP
Maximal value	2.525	2.333	3.877	3.607
Minimal value	0.0005	0.001	0.0006	0.0009
Average value	0.078	0.091	0.095	0.087
D DeWitt	1.487	0.181	1.087	**3.607**
M Stonebraker	**2.525**	0.632	0.481	**2.851**
C Faloutsos	0.357	0.242	**1.571**	**1.187**
W Bruce	0.538	**2.333**	0.172	0.483
R Agrawal	0.518	0.189	**3.877**	0.600
J Han	0.666	0.138	**2.029**	0.240

In order to show the influence results in more general areas, we select five categories of documents in Cora data and set the number of topics to five. Five meaningful topics according to the five categories: data mining ("DM"), information retrieval ("IR"), natural language processing ("NLP"), object-oriented database ("OODB") and database performance ("DBP") are obtained. Figure 3.16 shows several famous authors' estimated global influence distributions on the five topics. The results are very telling. For example, W Bruce is most influential on topic "IR," while R Agrawal and J Han are most influential on topic "DM." It is interesting to find that C Faloustsos is influential on both topic "DM" and topic "DBP," which is consistent to the real situation. Besides the two topics related to database, D DeWitt is also very influential on topic "DM." The reason should be that the area "DM" develops from database. Furthermore, Table 3.8 shows the maximal, minimal, and average values of the estimated global influence in the whole network w.r.t. each topic, which demonstrates that these authors almost have the largest values in their domains. Thus, it proves the validity of the way of global influence estimation.

Case Study 3: Topic-level influence comparison. Work [143] also proposed a method to discover topic-level influence. We compare the author influence results obtained by our model ($M1$) with the results by the model in [143] ($M3$). As sometimes it is hard to label the author influence

Table 3.9: Influencing author ranking w.r.t. several authors

D Blei		A McCallum		T Griffiths	
M1	M3	M1	M3	M1	M3
H Attias	D Blei	A McCallum	A McCallum	T Hofmann	T Griffiths
D Blei	M Stephens	D Blei	D Kauchak	M Steyvers	R Kass
M Jordan	J Pritchard	Andrew Ng	E Stephen	T Griffiths	N Chater
K Nigam	P Donnelly	T Griffiths	R Madsen	T Minka	D Lawson
T Jaakkola	C Meghini	M Jordan	C Elkan	A McCallum	H Neville

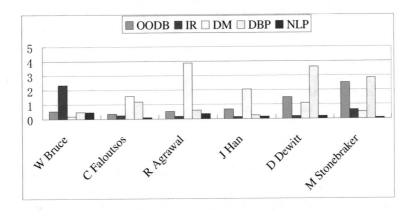

Figure 3.16: Estimated global influence distribution on topics.

strength, we only show the top five most influencing authors on some well-known researchers: Blei, McCallum, and Griffiths obtained by these two models in Table 3.9. The results demonstrate that our model can get meaningful results but $M3$ cannot. For example, our model discovers that Jordan, Blei, and Hofmann are one of the most influential researchers for Blei, McCallum, and Griffiths, respectively. But $M3$ does not get these results. As $M3$ only uses the link information of author citation, it will lose the information of relationships between authors and documents. Moreover, the assumption used in Tang et al. [143] which states that the node will be more influential if it has a great self-influence makes each person most influential on himself.

Similar to our model, $M3$ can also get the influence distributions on topics by inputting the nodes' topic mixtures. But the difference is that the topic information is used as an input prior instead of an integrated parameter in the method $M3$ while our method can obtain topics simultaneously. Figure 3.17 shows an example of the influence from Jordan to Blei and compares the topic distributions of influence obtained by our model and $M3$, respectively. First, Jordan and Blei's distributions on topics are illustrated, which indicate that both of them mainly work on Topic 3. Then, we can see that the influence obtained by our model has the largest strength on Topic 3 but the influence distribution from $M3$ is flat, from which it is not obvious to tell the influence semantic meaning. Thus, it is proved that our model can obtain more meaningful topic distributions of influence.

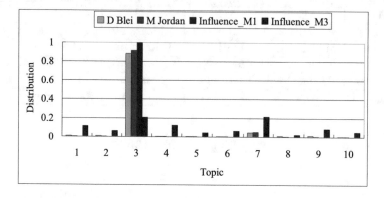

Figure 3.17: Topic distributions of authors and influence.

3.3.6 SUMMARY

In this section, we present an approach for mining topic-level influence in heterogeneous networks. The approach primarily consists of two steps, i.e., a probabilistic model to mine direct influence between nodes and different types of influence propagation methods to mine indirect and global influence. In the probabilistic model, we combine the textual content and heterogeneous link information into a unified generative process. Influence propagation methods further propagate influence along the links in the entire network. We also demonstrate that the learned influence can benefit several real prediction applications.

3.4 CONCLUSIONS

In this chapter, we introduce two methodologies for quantifying the topic-level influential strength between users for large social networks, and potential applications to user behavioral prediction.

CHAPTER 4

User Behavior Modeling and Prediction

We now turn to discuss how to leverage the different social phenomena to model and predict users' behaviors (social actions). From a broad viewpoint, the concept of user modeling is concerned with the process of building up a user model to characterize user's skills, declarative knowledge, and specific needs to a system [48]. We try to narrow down this concept in this book to modeling how users' behaviors (actions) are influenced by various factors such as personal interests, social influence, and global trends. Quite a few related studies have been conducted, for example, dynamic social network analysis [59, 82, 95, 125, 128], social influence analysis [2, 31, 42, 79, 121, 143], and group behavior analysis [5, 62, 130, 144, 149].

4.1 OVERVIEW

In sociology, the notion of social action was first proposed by Weber [159]. According to Weber, "an Action is 'social' if the acting individual takes account of the behavior of others and is thereby oriented in its course." We aim to systematically study how social actions evolve in a dynamic social network and to what extent different factors affect the user actions.

To clearly motivate this work, we conduct the following analysis on three real social networks: Twitter,[1] Flickr,[2] and ArnetMiner.[3] On Twitter, we define the action as whether a user discusses the topic "Haiti Earthquake" on his microblogs (tweets). On Flickr, we define the action as whether a user adds a photo to his favorite list. On ArnetMiner, the action is defined as whether a researcher publishes a paper on a specific conference (or journal). The analysis includes three aspects: (1) social influence; (2) time-dependency of users' actions; and (3) action correlation between users. Figure 4.1 shows the effect of social influence. We see that with the percentage of one's friends performing an action increasing, the likelihood that the user also performs the action is increased. For example, when the percentage of one's friends discussing "Haiti Earthquake" on their tweets increases, the likelihood that the user posts tweets about "Haiti Earthquake" is also increased significantly. Figure 4.2 illustrates how a user's action is dependent on his historic behaviors. It can be seen that a strong time-dependency exists for users' actions. For instance, on Twitter, users who posted tweets about "Haiti Earthquake" will have a much higher probability,

[1]http://www.twitter.com, a microblogging system.
[2]http://www.flickr.com, a photo sharing system.
[3]http://arnetminer.org, an academic search system.

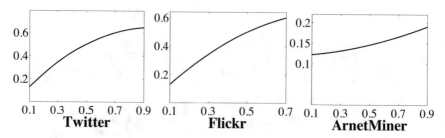

Figure 4.1:   Social influence. The x-axis stands for the percentage of one's friends who perform an action at $t-1$ and the y-axis represents the likelihood that the user also performs the action at t.

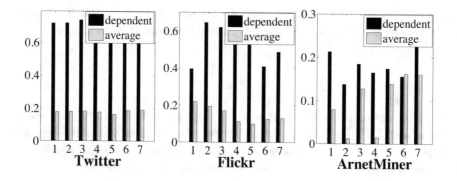

Figure 4.2:   Time-dependency of users' actions. The x-axis stands for different timestamps. "dependent" denotes the likelihood that a user performs an action which was previously performed by herself; "average" represents the likelihood that a user performs the action.

on average, (+20–40%) to post tweets on this topic than those who never discussed this topic on their blogs. Figure 4.3 shows the correlation between users' actions at the same timestamp. An interesting phenomenon is that friends may perform an action at the same time. E.g., on Twitter, two friends have a higher probability (+19.6%) to discuss "Haiti Earthquake" than two users randomly chosen from the network.

Problem Formulation.   Formally, we first give several necessary definitions and then formulate the problem of social action prediction.

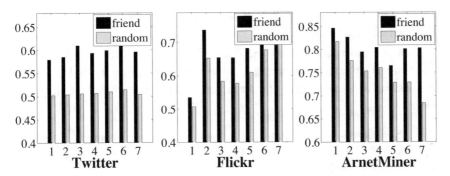

Figure 4.3: Action correlation. The x-axis stands for different time windows. "friend" denotes the likelihood that two friends perform an action together; "random" represents the likelihood that two random users perform the action together.

A static social network can be represented as $G = (V, E)$, where V is the set of $|V| = N$ users and $E \subset V \times V$ is the set of directed/undirected links between users. Given this, we can define the user's action as follows.

Definition 4.1   Action: An action y performed by user v_i at time t can be represented as a triple (y, v_i, t) (or shortly y_i^t). Let Y^t be the set of actions of all users at time t. Further, we denote all users' actions as the action history $\mathbf{Y} = \{(y, v_i, t)\}_{i,t}$.

Without loss of generality, we first consider the binary action, that is $y_i^t \in \{0, 1\}$, where $y_i^t = 1$ indicates that user v_i performed an action at time t, and $y_i^t = 0$ indicates that the user did not perform the action. Such an action log can be available from many online systems. For example, on Twitter, the action y_i^t can be defined as whether user v_i posts a tweet (microblog) about a specific topic (e.g., "Haiti Earthquake") at time t. Further, we assume that each user is associated with a number of attributes and thus have the following definition.

Definition 4.2   Time-varying attribute matrix: Let X^t be an $N \times d$ attribute matrix at time t in which every row $\mathbf{x}_i$ corresponds to a user, each column an attribute, and an element x_{ij} is the jth attribute value of user v_i.

The attribute matrix describes user-specific characteristics, and can be defined in different ways. For example, on Twitter, each attribute can be defined as a keyword and the value of an attribute can be defined as the frequency of a keyword occurring on a user's posted tweets. Thus, we can define the input of our problem, a set of attribute augmented networks.

Definition 4.3   Attribute augmented network: The attribute augmented network is denoted as $G^t = (V^t, E^t, X^t, Y^t)$, where V^t is the set of users and E^t is the set of links between users at time

t, and X^t represents the attribute matrix of all users in the network at time t, and Y^t represents the set of actions of all users at time t.

Based on the above concepts, we can define the problem of social action prediction. Given a series of T time-dependent attribute augmented networks, the goal is to learn a model that can best fit the relationships between the various factors and the user actions. More precisely, we have the following.

Problem 4.4 **Social action prediction.** Given a series of T time-dependent attribute augmented networks $\{G^t = (V^t, E^t, X^t, Y^t)\}$, where $t \in \{1, \cdots, T\}$, the goal of social action prediction is to learn a mapping function

$$f : (\{G^1, \ldots, G^{T-1}\}, V^T, E^T, X^T) \to Y^T.$$

Note that in this general formulation, we allow the graph structure to evolve over time and also arbitrary dependency from the past. To have a tractable problem to work with, we model the time-dependency by introducing a latent state for each user. More specifically, their actions are generated by their latent states, which are dependent on their neighbors' states at time t and $t - 1$.

Challenges and Solution. Thus, the problem becomes how to effectively and efficiently predict the dynamic users' actions. This problem is non-trivial and poses a set of unique challenges. First, the social network data (e.g., network structure and social actions) are very noisy. Users performing the same action may not have the same preference towards that action. Likewise, users who did not perform the action do not mean they have no interests towards the action. Second, user behaviors are highly time-dependent. For example, the influence of a user on another (strongly) depends on their historic interactions. Third, users' actions are usually correlated. In addition, as real social networks are getting larger with thousands or millions of users. It is important to develop the model that can scale well to real large data sets.

In this section, we try to systematically investigate the problem of social action prediction [138]. We officially formulate the problem of social action prediction and propose a unified model: Noise Tolerant Time-varying Factor Graph Model (NTT-FGM). We present an efficient algorithm for model learning and develop a distributed implementation based on MPI (Message-Passing Interface) to scale up to real large networks. We conduct experiments on three different data sets: Twitter, Flickr, and ArnetMiner. Experimental results show that the proposed NTT-FGM model can achieve a better performance for the action prediction than several alternative models.

4.2 APPROACH FRAMEWORK FOR SOCIAL ACTION PREDICTION

To summarize, for modeling and predicting social actions, we have the following intuitions:

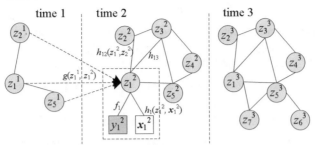

Figure 4.4: Graphical representation of the NTT-FGM model. Each circle stands for a user's latent action state z_i^t at time t in the network, which is used to characterize the intention degree of the user to perform the action; the latent state is associated with the action y_i^t, a vector of attributes $\mathbf{x}_i^t$, and depends on friends' historic actions $\mathbf{z}_{\sim v_i}^{t-1}$ and correlates with friends' actions $\mathbf{z}_{\sim v_i}^t$ at time t; $g(.)$ denotes a factor function to represent the friends' influence on a user's action; $h_i(.)$ represents a factor defined on user v_i's attributes; and $h_{ij}(.)$ represents a factor to capture the correlation between users' actions.

1. users' actions at time t are influenced by their friends' historic actions (time $< t$);

2. users' actions at time t are usually dependent on their previous actions; and

3. users' actions at a same time t have a (strong) correlation.

Moreover, the discrete variable y_i^t only models the user's action at a coarse level, but cannot describes the intention degree of the user to perform an action. Directly modeling the social actions Y would inevitably introduce noise to the model. Hence, a continuous variable for modeling the *action bias* is favorable.

With the intuitions discussed above, we propose a noise tolerant time-varying factor graph model (NTT-FGM) for social action prediction. Before explaining the model in detail, we first introduce the definition of latent action state.

Definition 4.5   Latent action state: For each user's action y_i^t, we define a (continuous) latent state $z_i^t \in [0, 1]$, which corresponds to a combination of the observed action y_i and a possible bias, to describe the actual intention degree of the user to perform the action.

Figure 4.4 shows the graphical structure of the NTT-FGM model. An action of user v_i at time t, i.e., y_i^t is modeled by using a (continuous) latent action state z_i^t, which is dependent on friends' historic actions $\mathbf{z}_{\sim v_i}^{t-1}$ (where $\sim v_i$ represents friends of user v_i in the network), users' action correlation $\mathbf{z}_{\sim v_i}^t$, and users' attributes $\mathbf{x}_i^t$. Specifically, in the NTT-FGM model, each discrete action is mapped into the latent state space and the action bias is modeled using a factor function. For example, for $y_i^t = 1$, a small value of its corresponding z_i^t suggests that a user v_i has a low intention to perform the action, thus a large action bias $|y_i^t - z_i^t|$. Next, influence between users is modeled using the latent states based on the same assumption as in HMM [52] and Kalman Filters [66]: latent states of users' actions at time t are conditionally independent of all the previous states given the latent states at time $t - 1$. Finally, actions' correlation is also modeled in the latent state space. A Markov random field is defined to model the dependency (correlation) among the continuous latent states. Different from the traditional Markov random field model (e.g., CRF [87], HMM [52], Kalman Filters [66]), the NTT-MRF model uses a continuous variable to describe the latent state, and utilizes a combination of multivariate Gaussian function and Markov random field to incorporate both time-inter and time-intra dependency between users' actions.

Now, we explain the proposed NTT-FGM model in detail. Given a series of attribute augmented networks $\mathbf{G} = \{G^t = (V^t, E^t, X^t, Y^t)\}$, $t \in \{1, \cdots, T\}$ and $V = V^1 \cup V^2 \cup \ldots \cup V^T$, $|V| = N$, we can define the joint distribution over the actions $\mathbf{Y}$ given $\mathbf{G}$ as

$$p(\mathbf{Y}|\mathbf{G}) = \prod_{t=1}^{T} \prod_{i=1}^{N} f(y_i^t|z_i^t) f(z_i^t|\mathbf{z}_{\sim v_i}^{t-1}) f(z_i^t|\mathbf{z}_{\sim v_i}^t, \mathbf{x}_i^t), \qquad (4.1)$$

where notation $\sim v_i$ represents neighbors of v_i in the social network. The joint probability has three types of factor functions, corresponding to the intuitions we have discussed. Specifically,

- action bias factor: $f(y_i^t|z_i^t)$ represents the posterior probability of user v_i's action y_i at time t given the continuous latent state z_i^t;

- influence factor: $f(z_i^t|\mathbf{z}_{\sim v_i}^{t-1})$ reflects friends' influence on user v_i's action at time t; and

- correlation factor: $f(z_i^t|\mathbf{z}_{\sim v_i}^t, \mathbf{x}_i^t)$ denotes the correlation between users' action at time t.

The three factors can be instantiated in different ways, reflecting our prior knowledge for different applications. In this chapter, we will give a general definition for the three factors. For the action bias factor $f(y_i^t|z_i^t)$, we define it using a Gaussian function:

$$f(y_i^t|z_i^t) = \frac{1}{\sqrt{2\pi\sigma^2}} \exp\{-\frac{(y_i^t - z_i^t)^2}{2\sigma^2}\}, \qquad (4.2)$$

where σ is a variance to control the bias and its value can be learned using an EM-style algorithm or predefined empirically. Note that if we only consider the binary action, the bias factor can be also defined based on a Bernoulli distribution.

For influence factor $f(z_i^t|\mathbf{z}_{\sim v_i}^{t-1})$, we first define an binary $N \times N$ matrix M^{t-1} to describe the social network at time $t - 1$, where the element $m_{ij}^{t-1} = 1$ represents that user v_i and v_j have a relationship in the social network (i.e., $e_{ij} \in E$), and $m_{ij}^{t-1} = 0$ indicates there is no relationship between v_i and v_j. Given this, we can formally define the influence factor as:

$$f(z_i^t|\mathbf{z}_{\sim v_i}^{t-1}) = \frac{1}{Z_1} \exp\{\sum_{j=1}^{N} \lambda_{ji} m_{ji}^{t-1} g_{ji}(z_i^t, z_j^{t-1})\}, \qquad (4.3)$$

where $g_{ji}(z_i^t, z_j^{t-1})$ is a function defined on the latent states of two users z_i^t and z_j^{t-1}; λ_{ji} (when $m_{ji}^{t-1} = 1$) represents the influence degree of v_j on v_i. For example, given a higher influence λ_{ji}, the action of user v_j is more likely to induce user v_i to behave in a similar way. Z_1 is a normalization factor. When $j = i$, we refer to the influence as self-influence, which actually characterizes the dependency of the user's action on his own previous state.

The correlation factor can be naturally modeled in a Markov random field. Therefore, by the fundamental theorem of random fields, we can define the correlation factor as:

$$
f(z_i^t | \mathbf{z}_{\sim v_i}^t, \mathbf{x}_i) = \frac{1}{Z_2} \exp\{ (\sum_{j=1}^{N} \beta_{ij} m_{ij}^t h_{ij}(z_i^t, z_j^t)
$$
$$
+ \sum_{k=1}^{d} \alpha_k h_k(z_i^t, x_{ik}^t)) \}, \tag{4.4}
$$

where $h_{ij}(z_i^t, z_j^t)$ is a feature function to capture the correlation between user v_i and v_j at time t; $h_k(z_i^t, x_{ik}^t)$ is a feature function defined on user v_i and the k-th attribute x_{ik}; d is the number of attributes; β_{ij} and α_k are, respectively, weights of the two functions; and Z_2 is again a normalization factor.

Finally, by integrating Equations (4.2)-(4.4) into Equation (4.1), we can obtain the following joint probability

$$
p(\mathbf{Y}|\mathbf{G}) = \frac{1}{Z} \exp\{ \sum_{t=1}^{T} \sum_{i=1}^{N} \frac{(y_i^t - z_i^t)^2}{2\sigma^2} + \sum_{t=1}^{T} \sum_{i=1}^{N} \sum_{j=1}^{N} \lambda_{ij} m_{ji}^{t-1} g(z_i^t, z_j^{t-1})
$$
$$
+ \sum_{t=1}^{T} \sum_{i=1}^{N} \sum_{j=1}^{N} \beta_{ij} m_{ij}^t h_{ij}(z_i^t, z_j^t) + \sum_{t=1}^{T} \sum_{i=1}^{N} \sum_{k=1}^{d} \alpha_k h_k(z_i^t, x_{ik}^t) \}, \tag{4.5}
$$

where $Z = (2\pi\sigma^2)^{\frac{N \times T}{2}} Z_1 Z_2$.

Learning NTT-FGM is to estimate a parameter configuration $\theta = (\{z_i\}, \{\alpha_k\}, \{\beta_{ij}\}, \{\lambda_{ij}\})$ from a given historic action log $\mathbf{Y}$, that maximizes the log-likelihood objective function $\mathcal{O}(\theta) = \log p_\theta(\mathbf{Y}|\mathbf{G})$, i.e.,

$$
\theta^\star = \arg \max \mathcal{O}(\theta). \tag{4.6}
$$

4.2.1 MODEL LEARNING

There are two challenges to solve the objective function. First, as the network structure in the social network can be arbitrary (may contain cycles), traditional methods such as Junction Tree [163] and Belief Propagation [170] cannot result in an exact solution. Second, to calculate the normalization factor Z, it is necessary to guarantee that the denominator of Equation (4.5), i.e., the exponential function $\exp\{.\}$, is integrable. Based on these considerations, we instantiate the factor functions $g(.)$ and $h(.)$ as follows:

$$g_{ji}(z_i^t, z_j^{t-1}) = -(z_i^t - z_j^{t-1})^2 \tag{4.7}$$
$$h_{ij}(z_i^t, z_j^t) = -(z_i^t - z_j^t)^2 \tag{4.8}$$
$$h_k(z_i^t, x_{ik}^t) = -(z_i^t - x_{ik}^t)^2. \tag{4.9}$$

We see that all of the factor functions are defined by quadratic functions. This is because quadratic equation satisfies the above two requirements: it is integrable and it offers the possibility to design an exact solution. Moreover, by defining in this way, the influence factor and the correlation factor can be elegantly explained with the information diffusion theory, by which the actions of users spread in the social network along the relationships [8, 59].

Finally, the objective function $\mathcal{O}(\theta)$ can be rewritten as

$$\mathcal{O}(\theta) = -\log Z - \{\sum_{t=1}^{T}\sum_{i=1}^{N}\frac{(y_i^t - z_i^t)^2}{2\sigma^2} + \sum_{t=1}^{T}\sum_{i=1}^{N}\sum_{j=1}^{N}\lambda_{ji}m_{ji}^{t-1}(z_i^t - z_j^{t-1})^2$$
$$+ \sum_{t=1}^{T}\sum_{i=1}^{N}\sum_{j=1}^{N}\beta_{ij}m_{ij}^t(z_i^t - z_j^t)^2 + \sum_{t=1}^{T}\sum_{i=1}^{N}\sum_{k=1}^{d}\alpha_k(z_i^t - x_{ik}^t)^2\} \tag{4.10}$$

where

$$Z = C\int_y\int_z\exp\{-\sum_{t=1}^{T}\sum_{i=1}^{N}\frac{(y_i^t - z_i^t)^2}{2\sigma^2} - \sum_{t=1}^{T}\sum_{i=1}^{N}\sum_{j=1}^{N}\lambda_{ji}m_{ji}^{t-1}(z_i^t - z_j^{t-1})^2$$
$$- \sum_{t=1}^{T}\sum_{i=1}^{N}\sum_{j=1}^{N}\beta_{ij}m_{ij}^t(z_i^t - z_j^t)^2 - \sum_{t=1}^{T}\sum_{i=1}^{N}\sum_{k=1}^{d}\alpha_k(z_i^t - x_{ik}^t)^2\}dzdy \tag{4.11}$$

where $C = (2\pi\sigma^2)^{\frac{N \times T}{2}}$ is a constant.

The Learning Algorithm.
The task of model learning is to estimate the parameters $\theta = (\{z_i\}, \{\alpha_k\}, \{\beta_{ij}\}, \{\lambda_{ij}\})$ by solving the objective function Equation (4.10). For this purpose, we need to first solve the integration of Z. As y is discrete, we can easily integrate out the first term in the $\exp\{.\}$ function of Equation (4.11). Furthermore, to guarantee that Z is integrable we must have $\alpha_k > 0, \beta_{ij} > 0, \lambda_{ij} > 0$. It is still difficult to solve the integration. To deal with this, our basic idea is to transform the exponential function $\exp\{.\}$ into a multivariate Gaussian distribution, and calculate the integration as follows:

$$Z = Const \cdot |A|^{-\frac{1}{2}} \exp\{\mathbf{b}^T A^{-1} \mathbf{b} - c\}, \tag{4.12}$$

where $c = \sum_{t=1}^{T} \sum_{i=1}^{N} \sum_{k=1}^{d} \alpha_k x_{ik}^t$; *Const* is a constant; A is a $NT \times NT$ block tridiagonal matrix; and $\mathbf{b} = \mathbf{X}\alpha$ is a NT-dimension vector and $\mathbf{X} = \{X^1 : X^2 : \cdots : X^T\}$ is $NT \times d$ matrix by concatenating all time-varying attribute matrices together.

Derivation of Z. We now give the details on how we obtain the integration of Z. Equation (4.11) can be rewritten in the form of a multivariate Gaussian distribution. The standard formation of the integration of Multivariate Gaussian Distribution is as follows:

$$\frac{1}{(2\pi)^{\frac{m}{2}} |M|} \int_{\mathbf{u}} \exp\{-\frac{1}{2}(\mathbf{u} - \boldsymbol{\mu})^T M^{-1}(\mathbf{u} - \boldsymbol{\mu})\} d\mathbf{u} = 1, \tag{4.13}$$

where $\mathbf{u}$ and $\boldsymbol{\mu}$ is a m-dimension vector, M is a $m \times m$ matrix.

The idea here is to transform the exponential function $\exp\{.\}$ in Equation (4.11) into a formation of multivariate Gaussian distribution.

$$\exp\{.\} \equiv \exp\{-\frac{1}{2}(\mathbf{z} - \boldsymbol{\mu})^T M^{-1}(\mathbf{z} - \boldsymbol{\mu}) - c\}, \tag{4.14}$$

where c is a value independent of $\mathbf{z}$. With further derivation, we can arrive

$$Z = Const \cdot |A|^{-\frac{1}{2}} \exp\{b^T A^{-1} b - c\} \tag{4.15}$$

where $\mathbf{b} = X\alpha$; $c = \sum_{t=1}^{T} \sum_{i=1}^{N} \sum_{k=1}^{d} \alpha_k x_{ik}^t$; A is a $NT \times NT$ block tridiagonal matrix, and $|A|$ is determinant of matrix A. The elements of A is defined as follows (we use i^t to denote $i + (t - 1) * N$ for simplicity):

$$A_{i^t, i^t} = \sum_{k=1}^{d} \alpha_k + \sum_{j=1}^{N} \beta_{ij} m_{ij}^t + \sum_{j=1}^{N} \beta_{ji} m_{ji}^t + \sum_{j=1}^{N} \lambda_{ji} m_{ji}^{t-1} + \sum_{j=1}^{N} \lambda_{ij} m_{ij}^{t+1}$$

$$A_{i^t, j^t} = A_{j^t, i^t} = -\beta_{ij} m_{ij}^t - \beta_{ji} m_{ji}^t$$

$$A_{i^t, j^{t-1}} = A_{j^t, i^{t-1}} = -\lambda_{ji} m_{ji}^{t-1} - \lambda_{ij} m_{ij}^{t-1}.$$

This construction matches our intuition. A_{i^t, i^t} represents the coefficient of $(z_i^t)^2$, while A_{i^t, j^t} represents the correlation factor, and $A_{i^t, j^{t-1}}$ describes the influence factor.

The Learning Algorithm and Its Derivation. Given this, we can design an EM-style algorithm to maximize $\mathcal{O}(\theta)$, as summarized in Algorithm 6:

- **E-step:** fix $\mathbf{z}$ and update all α, β, and λ, using a gradient descent method;

- **M-step:** fix α, β, and λ to update all $\mathbf{z}$, by solving a linear system.

More specifically, the algorithm for model learning primarily consists of two steps. To summarize, in the first step, we fix $\mathbf{z}$ and update $\boldsymbol{\alpha}, \boldsymbol{\beta}, \boldsymbol{\lambda}$ according to their gradients. We need to guarantee that $\alpha_k, \beta_{ij}, \lambda_{ij} > 0$. Thus, conventional gradient descent cannot be directly applied to the constrained problem. We employ a technique similar to that in Qin et al. [119]. Specifically we first maximize $\mathcal{O}(\theta)$ with respect to the log function. As a result, we get:

$$
\begin{aligned}
\nabla_{\log \alpha_k} &= -\alpha_k \left(\sum_{t=1}^{T} \sum_{i=1}^{N} (z_i^t - x_{ik}^t)^2 + \frac{\partial \log Z}{\partial \alpha_k} \right) \\
\nabla_{\log \beta_{ij}} &= -\beta_{ij} \left(\sum_{t=1}^{T} (z_i^t - z_j^t)^2 + \frac{\partial \log Z}{\partial \beta_{ij}} \right) \\
\nabla_{\log \lambda_{ij}} &= -\lambda_{ij} \left(\sum_{t=1}^{T} m_{ji}^{t-1} (z_i^t - z_j^{t-1})^2 + \frac{\partial \log Z}{\partial \lambda_{ij}} \right),
\end{aligned}
\tag{4.16}
$$

where

$$
\begin{aligned}
\frac{\partial \log Z}{\partial \alpha_k} &= -\frac{1}{2|A|} \frac{\partial |A|}{\partial \alpha_k} + \frac{\partial \vec{b}^T A^{-1} \vec{b}}{\partial \alpha_k} - \sum_{t=1}^{T} \sum_{i=1}^{N} x_{ik}^t{}^2 \\
&= -\frac{1}{2} (A^{-T}) :^T I : + X_{,k}^T A^{-1} \vec{b} - \vec{b}^T A^{-1} A^{-1} \vec{b} \\
&\quad + \vec{b}^T A^{-1} X_{,k} - \sum_{t=1}^{T} \sum_{i=1}^{N} x_{ik}^t{}^2 \\
\frac{\partial \log Z}{\partial \beta_{ij}} &= -\frac{1}{2|A|} \frac{\partial |A|}{\partial \beta_{ij}} + \frac{\partial \vec{b}^T A^{-1} \vec{b}}{\partial \beta_{ij}} \\
&= -\frac{1}{2} (A^{-T}) :^T \frac{\partial A}{\partial \beta_{ij}} : -\vec{b}^T A^{-1} \frac{\partial A}{\partial \beta_{ij}} A^{-1} \vec{b} \\
\frac{\partial \log Z}{\partial \lambda_{ij}} &= -\frac{1}{2|A|} \frac{\partial |A|}{\partial \lambda_{ij}} + \frac{\partial \vec{b}^T A^{-1} \vec{b}}{\partial \lambda_{ij}} \\
&= -\frac{1}{2} (A^{-T}) :^T \frac{\partial A}{\partial \lambda_{ij}} : -\vec{b}^T A^{-1} \frac{\partial A}{\partial \lambda_{ij}} A^{-1} \vec{b},
\end{aligned}
\tag{4.17}
$$

where the notation $M :$ with a colon denotes the long column vector formed by concatenating the columns of matrix M.

In the second step, we fix $\boldsymbol{\alpha}, \boldsymbol{\beta}, \boldsymbol{\lambda}$ to update $\vec{z}$, by solving a linear system:

$$
(A + \mathbf{I})\vec{z} = \vec{y} + X\vec{\alpha}.
\tag{4.18}
$$

Social Action Prediction. Based on the learned parameters θ, we can predict the users' future actions. Specifically, for predicting a user's action y_i^{T+1} at time $T + 1$, we first compute the latent state z_i^{T+1}; and then use the latent state to infer the action y_i^{T+1}. To compute the latent state z_i^{T+1}, we have the following formula:

Input: number of iterations I and learning rate η;
Output: learned parameters $\theta = (\{z_i\}, \{\alpha_k\}, \{\beta_{ij}\}, \{\lambda_{ij}\})$;

Initialize $\mathbf{z} = \mathbf{y}$;
Initialize α, β, λ;
repeat
    E Step: % fix $\mathbf{z}$, learn α, β, λ;
    for $i = 1$ *to* I **do**
        Compute gradient $\nabla_{\log \alpha_k}, \nabla_{\log \beta_{ij}}, \nabla_{\log \lambda_{ij}}$;
        Update $\log \alpha_k = \log \alpha_k + \eta \times \nabla_{\log \alpha_k}$;
        Update $\log \beta_{ij} = \log \beta_{ij} + \eta \times \nabla_{\log \beta_{ij}}$;
        Update $\log \lambda_{ij} = \log \lambda_{ij} + \eta \times \nabla_{\log \lambda_{ij}}$;
    end
    M Step: % fix α, β, λ learn $\mathbf{z}$;
    Solve the following linear equation:

$$(A + \mathbf{I})\mathbf{z} = \mathbf{y} + X\alpha$$

until *convergence*;

Algorithm 6: Expectation maximization.

$$z_i^{T+1} = \frac{\sum_{k=1}^{d} \alpha_k x_{ik} + \sum_{j=1}^{N} \lambda_{ji} m_{ji}^T z_j^T}{\sum_{k=1}^{d} \alpha_k + \sum_{j=1}^{N} \lambda_{ji} m_{ji}^T}. \tag{4.19}$$

However, the above equation calculates the latent state independently and ignore the correlation between actions. By further considering the action correlation factor, that is to compute all $\mathbf{z}$ together, we can solve the following linear system:

$$\forall i, \quad \sum_{k=1}^{d} \alpha_k (z_i^{T+1} - x_{ik}) + \sum_{j=1}^{N} \lambda_{ji} m_{ji}^T (z_i^{T+1} - z_j^T)$$
$$+ \sum_{j=1}^{N} \beta_{ij} (z_i^{T+1} - z_j^{T+1}) + \sum_{j=1}^{N} \beta_{ji} (z_i^{T+1} - z_j^{T+1}) = 0. \tag{4.20}$$

Then, we can predict the users' actions y according to their corresponding latent states z by:

$$y_i^{T+1} = \begin{cases} 0 & \text{if } |z_i^{T+1} - \bar{z}_+| <= |z_i^{T+1} - \bar{z}_-| \\ 1 & \text{otherwise.} \end{cases}, \tag{4.21}$$

where $\bar{z}_+$ and $\bar{z}_-$ are, respectively, the average state values of the corresponding actions $y = 1$ and $y = 0$ in the training data, and are computed by:

$$\bar{z}_- = \frac{\sum_{t=1}^{T} \sum_{i=1}^{N} z_i^t I(y_i^t = 0)}{\sum_{t=1}^{T} \sum_{i=1}^{N} I(y_i^t = 0)} \qquad (4.22)$$

$$\bar{z}_+ = \frac{\sum_{t=1}^{T} \sum_{i=1}^{N} z_i^t I(y_i^t = 1)}{\sum_{t=1}^{T} \sum_{i=1}^{N} I(y_i^t = 1)} \qquad (4.23)$$

where I is the indicator function.

Distributed NTT-FGM Learning. As a social network may contain millions of users and hundreds of millions of social ties between users, it is impractical to learn a NTT-FGM from a huge data using a single machine. Specifically, there are two major problems in our NTT-FGM model, namely, memory space and computing time. We use a sparse representation to solve the first problem. To speed up the computing, we deploy the learning task on a distributed system based on the MPI (Message Passing Interface).

MPI is a message-passing library interface specification. In the message-passing parallel programming model, data is moved from the address space of one process to that of another process through cooperative operations on each process. Based on the message passing scheme, we employ the *master-slave* model. That is, master can assign tasks to the slaves (computers), and combine the results in the master machine.

Specifically, in our learning algorithm, the time-consuming step lies in the calculation of the gradients, $\nabla_{\log \alpha_k}$, $\nabla_{\log \beta_{ij}}$, $\nabla_{\log \alpha_{ij}}$, which requires computing the inverse of the matrix A. Note A is a $NT \times NT$ matrix, which is too large to be held in memory when deal with a large data. Thus, we compute each column of A^{-1} respectively by solving the following linear equation

$$\forall i \quad Ax_i = b_i, \qquad (4.24)$$

where x_i represents the i column of A^{-1} and b_i represents a NT-dimension vector, with the ith element 1, the other elements 0. Thus in each iteration, the master broadcasts the parameters to each slave and assigns the tasks to solve Equation (4.24) to the slaves averagely. All the salve computers calculate A, and send the results back to the master. The master reduces all the distributed results, and broadcasts the updated parameters to the slaves again for the next iteration. The detailed description is in Algorithm 7.

4.3   EVALUATION

The proposed approach for social action prediction is very general and can be applied to analyze different kinds of social networks. In this section, we present various experiments to evaluate the effectiveness and efficiency of the proposed approach. All data sets and codes are publicly available.[4]

[4]http://arnetminer.org/stnt/

Input: number of iterations I and learning rate η;
Output: learned parameters $\theta = (\{z_i\}, \{\alpha_k\}, \{\beta_{ij}\}, \{\lambda_{ij}\})$;

Initialize $\mathbf{z} = \mathbf{y}$;
Initialize α, β, λ;
repeat
    Master broadcasts $\mathbf{z}$;
    E Step: % fix $\mathbf{z}$, learn α, β, λ;
    for $i = 1$ *to* I **do**
        Master broadcasts α, β, λ;
        Compute gradient $\nabla_{\log \alpha_k}, \nabla_{\log \beta_{ij}}, \nabla_{\log \lambda_{ij}}$ each;
        Slaves send back the calculation results;
        Master reduces the results;
        Master Update α, β, λ;
    end
    M Step: % fix α, β, λ learn $\mathbf{z}$;
    Master Solve the following linear equation:

$$(A + \mathbf{I})\mathbf{z} = \mathbf{y} + X\alpha$$

until *convergence*;

Algorithm 7: Parallel expectation maximization.

4.3.1 EVALUATION METRICS

Data Sets We perform our experiments on three different genres of real-world data sets: Twitter (a microblogging data set crawled from twitter.com), Flickr (a data set of photo sharing from flickr.com), and ArnetMiner (a publication data set from arnetminer.org).

- Twitter. The data set is crawled from Twitter by starting from the user "Carel Pedre (carelpedre),"[5] one of Haitian most popular radio DJs, who used Twitter to inform the world about the earthquake which ravaged his country. We extract all followers ($> 11,704$) of "carelpedre" and the users he is following, and continue the process for each extracted Twitter user. We further crawl all tweets posted by the users as attributes. Finally, a data set used for action prediction consists of 7,521 users, 304,275 time varying following and followed relationships, and 730,568 tweets (blogs) posted by the users. A larger data set consisting of millions of users is also publicly available.

- Flickr. The data set is collected by Cha et al. [25], which contains 8,721 users, 485,253 friendships between users, and 2,504,849 favorite photos .

- ArnetMiner. It is collected from ArnetMiner [148] and consists of 640,134 researchers, 1,554,643 coauthor relationship, and 2,329,760 publication papers by the researchers.

[5]http://www.carelpedre.com/

The action in Twitter is defined as the topic (e.g., "Haiti Earthquake") discussed by the user. More specifically, we selected several very relevant keywords, e.g., "Haiti," "earthquake," and "rescue." If a user posts a tweet containing the topic (keyword), we say that the user performs the action. We crawled the data from January 12th, when the Haiti Earthquake struck, to January 26th. In the Twitter data, we view one day as a time stamp. For example, if a user calls for a donation for Haiti, his friends may respond by re-tweeting it or posting a supporting message.

For Flickr, however, the action data is defined as whether or not a user adds a photo to his favorite list. For example, if a user added a photo to his favorite list, his friends may also add the photo to their favorites. We extract the historic action log from 11/1/2006 to 3/20/2007 in the data set, dividing into 14 time stamps, 10 days a stamp.

The action of the ArnetMiner data is defined as whether a researcher publishes a paper at a specific venue. For example, if a researcher published a paper at KDD, it may influence his collaborators to publish papers at KDD as well. The data is split into 10 time stamps, 1 for each year.

On all the three data sets, the attributes X is defined as the contents of tweets, information of photos, or related publication venues of the researcher. The content of each tweet is preprocessed by (a) removing stop-words and numbers; (b) removing words that appear less than three times in the corpus; and (c) lowercasing the obtained words. Then for each user we combine all words in the remaining words in the tweets posted by the user and create the attribute vector by taking words as features.

Evaluation Metrics. To evaluate our method, we consider the following three angles.

- **Prediction.** We evaluate the proposed model in terms of Precision, Recall, and F1-Measure, and compare with the baseline methods to validate the effectiveness of the proposed model.

- **CPU time.** It is the execution elapsed time of the model learning. This shows the speedup of the parallel implementation.

- **Case study.** We use several case studies as the anecdotal evidence to further demonstrate the effectiveness of our method.

We compare the following methods for social action prediction.

SVM: Utilizes users' associated attributes as well as their neighbors' states to train a classification model and then employs the classification model to predict users' actions. For SVM, we employ SVM-light.[6]

wvRN: Employs a weighted-vote relational neighbor classifier to train a classification model by making use of network information. In prediction, the relational classifier estimates the action state of a user by the weighted mean of his neighbors.

NTT-FGM: Uses the proposed NTT-FGM model to train the action tracking model and further uses the learned model for prediction.

---

[6]http://svmlight.joachims.org/

Table 4.1: Performance of action prediction with different approaches (%)

Data set	Method	Recall	Precision	F1-Measure
Twitter	SVM	10.41	16.71	13.85
	wvRN	0.45	7.89	0.86
	NTT-FGM	26.40	21.14	23.47
Flickr	SVM	34.48	45.05	39.06
	wvRN	60.02	48.81	53.84
	NTT-FGM	56.18	45.80	50.47
ArnetMiner	SVM	10.19	21.62	13.85
	wvRN	14.83	16.39	15.57
	NTT-FGM	31.14	44.28	36.57

According to our preliminary experiments, the σ in the Gaussian distribution does not significantly influence the performance. Thus, for simplicity, we empirically set $\sigma = 1$.

All the algorithms are implemented using C++ and all experiments are performed on a server running Ubuntu 8.10 with a AMD Phenom(tm) 9650 Quad-Core Processor (2.3 GHz) and 8 GB memory. The distributed learning algorithm is implemented under the MPI parallel programming model.[7] We perform the distributed training on 5 computer nodes (20 CPU cores) with AMD processors (2.3 GHz) and 40 GB memory in total. We set the maximum number of iterations as 250 and the threshold for the change of α, β, and λ to $1e - 3$.

4.3.2 PREDICTION PERFORMANCE

On all the three data sets, we use the historic users' actions to train the action tracking model and use the learned model to predict the users' actions in the last time stamp.

Table 4.1 lists the prediction performance of the different approaches on the three data sets with the following observations.

Performance comparison Our method NTT-FGM consistently achieves better performance comparing to the baseline methods. In terms of F1-Measure, NTT-FGM can achieve a +10% improvement compared with the (SVM). At the same time, NTT-FGM gives robust results, while the performance of wvRN is very sensitive to the data characteristics, with the highest F1-Measure on the Flickr data and extremely low value in the Twitter data. This is because on Flickr the user's action of adding favorite photos is mainly influenced by her friends' actions and wvRN can be viewed as a simple influence model, which makes wvRN mostly predicts "1" on Flickr, but the Twitter network (about "Haiti earthquake") in our experiment is relatively sparse, as a result

[7]http://www.mcs.anl.gov/research/projects/mpich2/

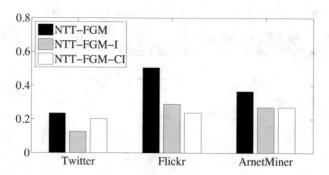

Figure 4.5: Contribution of different factor functions. NTT-FGM-I stands for our method by ignoring the influence factor function ($\lambda = 0$); and NTT-FGM-IC stands for NTT-FGM by ignoring both influence factor and correlation factor ($\lambda = 0$, $\beta = 0$).

wvRN outputs all "0." While our approach shows robust and consistent performance on all the data sets, which is important for the extendability of the methods.

Factor contribution analysis NTT-FGM captures three factors: (1) influence, (2) correlation, and (3) personal interests/attributes. Next we perform an analysis to evaluate the contribution of different factors defined in our model. In particular, we remove those factors one by one (first influence factor function, followed by the correlation factor function), and then train and evaluate the prediction performance of NTT-FGM. Figure 4.5 shows the F1-Measure score after ignoring the factor functions. We can observe clear drop on the prediction performance, which indicates that our method works well by integrating the different factors for action tracking (prediction) and each defined factor in our method contributes improvement in the performance. Also, we find that the decrease varies on different data sets. On Twitter there is a very low correlation between users' actions because users mainly post tweets on Twitter based on their previous experience or friends' tweets, and relatively act independently at a same time t.

Latent action states The learned latent action states essentially play a role as smoothing. Figure 4.6 illustrates several examples of the learned latent action states. It can be easily seen that the learned latent states (denoted as the red curve) is much more smoothing than the original discrete actions (denoted as the black step line), which indicates that latent action states can model the bias in binary actions. This is desirable for most prediction/classification tasks and further confirms us the advantage of the proposed NTT-FGM model.

4.3.3   EFFICIENCY PERFORMANCE

We now evaluate the efficiency of our approach by comparing the distributed learning algorithm with the basic one on the three data sets.

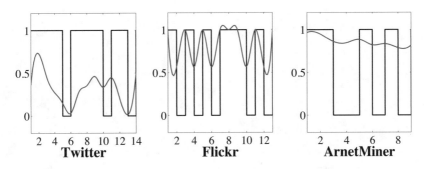

Figure 4.6:   Example latent action states.

Table 4.2 lists the CPU time required for learning the NTT-FGM model on a single machine (Basic NTT-FGM) and by the distributed learning algorithm using five computer nodes (each four cores). The distributed learning algorithm typically achieves a significant reduction of the CPU time. For example, on ArnetMiner, we obtain a speedup> 17×, and on Flickr, the distributed learning algorithm results in a speedup> 13×.

Table 4.2:   Efficiency performance on the three data sets (five computer nodes, each four cores)

Data Set	Basic NTT-FGM	Distributed NTT-FGM
Twitter	77.7hr	7.0hr
Flickr	9.14hr	0.68hr
ArnetMiner	100min	6.2min

We also evaluate the speedup of the distributed learning algorithm using different numbers of computer nodes (5, 10, 15, 20 cores) to evaluate the cost of message passing. The speedup, as shown in Figure 4.7(a), is close to the perfect line in the beginning. Although it decreases inevitably as the number of cores increases, it scales very well with > 10× speedup using 15 threads.

We further analyze how the network structure affects the efficiency of the learning algorithm. We generate a synthetic data set for this experiments by varying the density of the network $(\log \frac{|E|}{|V|})$. It can be seen from Figure 4.7(b) and (c) that as the density (x-axis) increases, both basic learning and the distributed learning algorithm need more CPU time to train the NTT-FGM model, but the speedup of the distributed algorithm is consistently high (about $14 - 15$× using 20 threads).

4.3.4   QUALITATIVE CASE STUDY

Now we present three case studies to demonstrate the effectiveness of the proposed model.

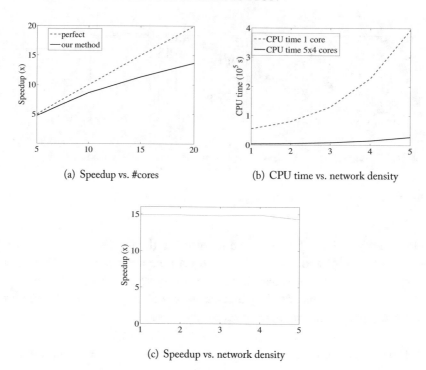

(a) Speedup vs. #cores

(b) CPU time vs. network density

(c) Speedup vs. network density

Figure 4.7: Speedup results. In graph (a), we evaluate the speedup varied with the number of cores. The x-axis stands for the number of cores, the y-axis represents the speedup($\frac{\text{CPU time 1 core}}{\text{CPU time } n \text{ cores}}$). In graph (b), we evaluate the CPU time with different network density(defined as $\log\frac{|E|}{|V|}$). The x-axis is $\log\frac{|E|}{|V|}$, the y-axis is the running time in seconds. In graph (c), we evaluate the speedup with different network density. The x-axis is $\log\frac{|E|}{|V|}$, the y-axis is the speedup.

"Haiti Earthquake" The Haiti Earthquake was a devastating earthquake, leaving the country in shambles. We use our results to analyze people's actions related to the catastrophe on Twitter. Table 4.3 lists several example tweets about the "Haiti Earthquake." We see that these tweets are about a call-for-donation by the famous tennis player "Serena Williams (serenajwilliams) ." The call-for message was soon retweeted by "actsofFaithblog" and "madameali" on their own microblogs, and a bit later the Haitian radio host "carelpedre" added a comment on Serena Williams's Twitter. These Twitter users are one of the most influential users and their actions on "Haiti Earthquake" quickly spread on Twitter with retweets and replies. (Because of this, Carel Pedre received a special "humanitarian" award at the second annual "Shorty Awards" in New York.) With the proposed model, we can identify the most influential users, whose actions can induce a large cascade followings, and track the information flows (via social ties with a high influence score or correlation score). In this way, we can understand how the influence spreads among people.

Table 4.3: Action tracking on Twitter for "Haiti Earthquake"

Date/User	Tweet
6:03 PM Jan 16th by extratv	Tennis pro Roger Federer is joining forces with Rafael Nadal & @serenajwilliams to raise money for Haiti. http://su.pr/1E3MDU
5:23 AM Jan 17th by serenajwilliams	Hey. Please, check out my foundation website: www.theswf.org to help those in Haiti!
6:48 AM Jan 17th by madameali	RT @SIXTWELVEMAG: RT @serenajwilliams: Hey. Please, check out my foundation website: www.theswf.org to help those in Haiti!
7:34 AM Jan 17th by actsofFaithblog	RT @serenajwilliams: Hey. Please, check out my foundation website: www.theswf.org to help those in Haiti!
2:50 PM Jan 17th by carelpedre	@serenajwilliams Through Her 92k Mission has set a goal to contribute donations to the victims in #haiti. Visit www.theswf.org and donate

Table 4.4: Prediction on who will publish on (or submit to) KDD 2010. The examples are selected from the top 100 researchers predicted by the NTT-FGM model

Frequent	Jiawei Han	Christos Faloutsos	Philip S. Yu
	Pedro Domingos	Lise Getoor	Jon M. Kleinberg
	Hang Li	ChengXiang Zhai	Wei-Ying Ma
	Lise Getoor	Jure Leskovec	Qiaozhu Mei
	Bing Liu	Jian Pei	Ravi Kumar
New	Huijia Zhu	Dimitrios Kotsakos	Zi Yang
	Noman Mohammed	Caimei Lu	Quanquan Gu
	Zhili Guo		

"Publication at KDD" We can also use the NTT-FGM model to track and predict who will publish (or submit) papers to KDD 2010. We train the NTT-FGM model using the ArnetMiner data before 2009 and use the learned model to predict the latent action state of each researcher, and finally obtain a list of researchers ranked by the latent state. Table 4.4 lists a few representative examples selected from the top 100 ranked researchers. We see that our approach cannot only find some famous researchers but also discover some "newcomers" to the KDD community. The first row lists several well-established researchers who have published a lot on KDD. The second row shows several "new" researchers who have no paper (or only few papers) published at KDD.

"Correlation between Researchers" Based on the learned NTT-FGM model, we can generate a correlation/influence map for better user analysis. Figure 4.8 shows an example correlation map between researchers. The strength of the link between two researchers indicates the correlation score. We see some researchers have strong correlation because they coauthored quite a few papers, e.g., Jiawei Han and Philip Yu. While our approach also finds some researchers have strong correlation, e.g., Ravi Kumar and Christos Faloutsos, although they only

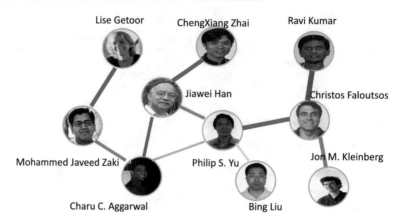

Figure 4.8: Example correlation analysis between researchers. The strength represents the correlation score between two researchers.

coauthored one or two paper(s). The discovered correlation can potentially benefit many applications such as link prediction. More correlation/influence analysis results can be found at `http://arnetminer.org/stnt/`.

4.4   SUMMARY

In this chapter, we study an interesting problem of user behavior modeling and prediction in dynamic networks. We propose a noise tolerant time-varying factor graph model (NTT-FGM) to formalize this problem in a unified model. Three factor functions are defined to capture the intuitions discovered in our observation and an efficient algorithm is presented to learn the tracking model. A distributed learning algorithm has been implemented under the message-passing parallel programming model. We experiment on three different genres of data sets and further present a case study on social action prediction using the learned NTT-FGM model. Experimental results on three different types of data sets demonstrate that the proposed approach can effectively model the social actions and clearly outperforms several alternative methods for action prediction. The distributed learning algorithm also has a good scalability performance.

CHAPTER 5

ArnetMiner: Deep Mining for Academic Social Networks

In this chapter, we use an online system, ArnetMiner [148],[1] to explain how the technologies presented in previous sections help real applications. ArnetMiner is a system aiming to extract and deep analyze academic social networks. Specifically, it provides four key functions: (1) extraction of a researcher social network automatically from the existing Web; (2) integration of the publications into the social network from existing digital libraries; (3) modeling of the whole academic social network; and (4) expertise oriented search using social network. The system has been in operation since 2006. So far, it has extracted information of more than 31,222,410 researchers and 69,962,333 publications from the Internet. The system has attracted 5,520,000 independent IP accesses from 220 countries (and regions) in the world up to 2014.

5.1   OVERVIEW

Previously, several issues in academic social network have been investigated and systems were developed (e.g., Microsoft Academic Search,[2] Rexa.info,[3] and Google Scholar[4]). However, most of the problems are investigated separately and the methods proposed are not sufficient for mining the whole academic social network. This is because of two reasons. (1) Lack of semantic-based information. The social information obtained solely from the user entered profile or extracted by using heuristics is sometimes incomplete or inconsistent. Users do not fill some information merely because they are unwilling to fill the information. (2) Lack of a unified modeling approach for efficient mining the social network. Traditionally, different typed information sources in the academic social network were modeled individually, and thus dependencies between them cannot be captured. However, dependencies (even strong dependencies) exist between the social data. High quality search services need to consider the intrinsic dependencies between the different information sources.

In ArnetMiner, we try to address the above challenges in novel approaches. Our objective in this system is to answer four questions: (1) How does one automatically extract/create the researcher profile from the existing Web? (2) How does one integrate the extracted information

[1]http://aminer.org
[2]http://academic.research.microsoft.com/
[3]http://rexa.info
[4]http://scholar.google.com

What's new?

AMiner II is the second generation of ArnetMiner, providing deep analysis and mining for scientific data:

- **Data Fusion:** Integrate scientific data from multiple sources.
- **Profiling:** Automatically create profile for each researcher, including basic information, research interest, social circles, and publication records.
- **Expert finding:** Find right experts, rising stars, reviewers, and collaborators.
- **Academic meter:** Measure scientific impact of authors, journals, and papers, and estimate the contribution of each collaborator.
- **Knowledge trend:** Identify the evolutionary trend from the scientific data, predict future development.
- **ArnetApp platform:** Use RESTFul API to invoke AMiner's services, deploy your applications on AMiner and share them with researchers all around the world.

ArnetMiner since 2006

User distribution

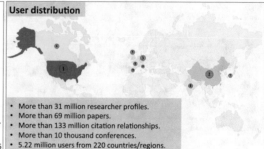

- More than 31 million researcher profiles.
- More than 69 million papers.
- More than 133 million citation relationships.
- More than 10 thousand conferences.
- 5.22 million users from 220 countries/regions.

Figure 5.1: ArnetMiner overview. The left figure lists the major functionalities in ArnetMiner and the right figure shows user distributions in the world.

(i.e., researchers' profiles and publications) from different sources? (3) How does one model the different typed information sources in a unified model? (4) How does one provide powered search services in the constructed network?

For the first question, we extend FOAF (Friend-Of-A-Friend) ontology [19] as the schema and employ a unified approach based on Conditional Random Fields (CRFs) to extract the profile of a researcher from the Web. For the second question, we integrate the extracted researcher profiles and publications from online digital libraries. We propose a unified probabilistic framework for dealing with the name ambiguity problem in the integration. For the third question, we propose three generative probabilistic models for simultaneously modeling characteristics of document contents, author interests, and conference themes. For the last question, we, based on the modeling results, propose several methods for expertise search (searching expertise authors, conference/journals, and papers), association search, author interests finding, and academic suggestion.

Architecture. Figure 5.2 shows the architecture of the system. The system mainly consists of five main components.

1. *Extraction*: Focuses on automatically extracting the researcher profile from the Web. It first collects and identifies one's relevant pages (e.g., homepages or introducing pages) from the Web, then uses a unified approach to extract data from the identified documents. It also extracts publications from online digital libraries using heuristic rules.

2. *Integration*: Integrates the extracted researchers' profiles and the extracted publications. It employs the researcher name as the identifier. A probabilistic framework has been proposed to deal with the name ambiguity problem in the integration. The integrated data is stored into a researcher network knowledge base (RNKB).

3. *Storage and Access*: Provides storage and indexing for the extracted/integrated data in the RNKB. Specifically, for storage it employs Jena [24], a tool to store and retrieve ontological

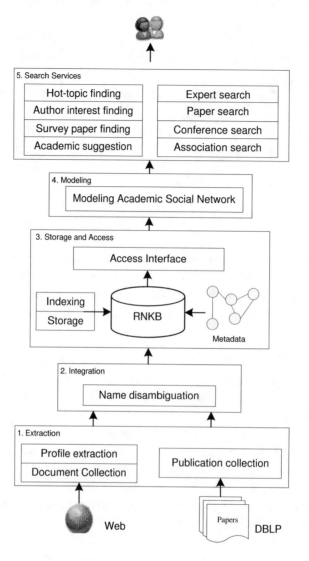

Figure 5.2: Architecture of ArnetMiner.

data; for indexing, it employs the inverted file indexing method, an existing method in information retrieval [153].

4. *Modeling*: Utilizes a generative probabilistic model to simultaneously model the different typed information sources. It estimates a mixture topic distribution associated with the different information sources.

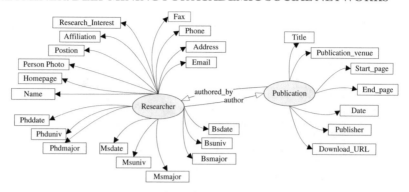

Figure 5.3: The schema of the researcher profile.

5. *Search Services*: Provides several powered search services based on the modeling results: expertise author search, paper search, conference search, and people association search. It also provides other services for supporting advanced applications, including: hot-topic finding, author interesting finding, survey paper finding, and academic suggestion.

For several features in the system, e.g., extraction of researchers' profiles, name disambiguation in the integration, academic modeling, and several search services (i.e., expertise search and association search), we propose new approaches trying to overcome the drawbacks that exist in the conventional methods. For some other features, e.g., storage and access, we utilize the state-of-the-art methods. This is because, these issues have been intensively investigated previously and the conventional methods can result in good performances. In the rest of the book, we will introduce in detail the challenges we are dealing with and describe our methods.

We first explain our approach to researcher profiling, and then describe the probabilistic framework to name disambiguation. Next, we propose three generative probabilistic models to model the constructed academic social network, and present several search services provided in ArnetMiner based on the modeling results.

5.2 RESEARCHER PROFILE EXTRACTION

We define the schema of the researcher profile (as shown in Figure 5.3), by extending the FOAF ontology [19]. In the profile, 24 properties and two relations are defined [145, 146].

It is non-trivial to perform the researcher network extraction from the Web. We produced statistics on randomly selected 1, 000 researchers. We observed that 85.62% of the researchers are faculties from universities and 14.38% are from company research centers. For researchers from the same company, they often have a template-based homepage. However, different companies have absolutely different templates. For researchers from universities, the layout and content of the homepages vary largely depending on the authors. We have also found that 71.88% of the

Table 5.1:   Content features, pattern features, and term features

		Content Feature		Pattern Feature
				All Token
Standard Token	Word	Word in the token	Positive word	If the token contains a pre-defined positive word
	Morphology	Morphology of the word	Negative word	If the token contains a pre-defined negative word
	Size	The size of the image	Special token	If the token contains a special pattern
	Height/width ratio	the ratio of height/width of the image	Name	If the token contains the researcher name
Image Token	Image format	The format of the image (e.g., "JPG")	#Line break	How many line breaks before
	Image color	The number of "unique color"		**Term Feature**
		The number of bits for per pixel		**Term Token**
	Filename	Words in the filename	Term	If the token contains a base noun phrase
	Face detection	If the image contains a person face recognized by (opencvlibrary.sf.net)		
	ALT	Words in "alt" attribute of the image	Dictionary	If the token contains a word in a dictionary

$1,000$ web pages are researchers' homepages and the rest are pages introducing the researchers. Characteristics of the two types of pages significantly differ from each other.

We analyzed the content of the web pages and found that about 40% of the profile properties are presented in tables or lists and about 60% are presented in natural language text. This also means a method without using the global context information in the page would be ineffective. A statistical study also unveils that (strong) dependencies exist between different profile properties. For example, there are $1,325$ cases (14.54%) in our data that extraction of the property needs to use the extraction results of the other properties. An ideal method should consider processing all the subtasks together.

5.2.1   A UNIFIED APPROACH TO PROFILING

The proposed approach consists of three steps: relevant page identification, preprocessing, and tagging. In relevant page identification, given a researcher name, we first get a list of web pages by a search engine (we used Google API) and then identify the homepage/introducing page using a classifier. We use Support Vector Machines (SVM) [30] as the classification model and define features such as whether the title of the page contains the person name and whether the URL address (partly) contains the person name. The performance of the classifier is 92.39% in terms of F1-measure. In preprocessing, (A) we separate the text into tokens and (B) we assign possible tags to each token. The tokens form the basic units and the pages form the sequences of units in the tagging problem. In tagging, given a sequence of units, we determine the most likely corresponding sequence of tags by using a trained tagging model. (The type of the tags corresponds to the property defined in Figure 5.3.) In this chapter, as the tagging model, we make use of Conditional Random Fields (CRFs) [87]. Next, we describe the steps (A) and (B) in detail.

(A) We identify tokens in the Web page using heuristics. We define five types of tokens: *standard word*, *special word*, *image* token, term, and punctuation mark. Standard words are unigram words in natural language. Special words [135] include email, URL, date, number, percentage, words containing special symbols (e.g., "Ph.D." and ".NET"), unnecessary tokens (e.g., "===" and "###"), etc. We identify special words by using regular expressions. *image* tokens are *image* tags in the HTML file. We identify it by parsing the HTML file. Terms are base noun phrases extracted from the web pages. We employed a tool based on technologies proposed in Xun et al. [166].

(B) We assign tags to each token based on the token type. For example, for standard word, we assign all possible tags (each tag representing a property). For special word, we assign tags: Position, Affiliation, Email, Address, Phone, Fax, Bsdate, Msdate, and Phddate. For *image* token, we assign two tags: Photo and Email (an email is likely to be shown as an image).

In this way, each token can be assigned with several possible tags. Using the tags, we can perform most of the profiling processing (conducting 16 subtasks defined in Figure 5.3).

The CRF Model.   We employ Conditional Random Fields (CRF) as the tagging model. CRF is a conditional probability of a sequence of labels Y given a sequence of observations tokens X,

Table 5.2: Performances of researcher profiling (%)

Profile	Unified			Unified_NT			SVM			Amilcare		
	Prec.	Rec.	F1	Prec.	Rec.	F1	Prec.	Rec.	F1	Prec.	Rec.	F1
Photo	90.32	88.09	89.11	89.22	88.19	88.64	87.99	89.98	88.86	97.44	52.05	67.86
Position	77.53	63.01	69.44	73.99	57.67	64.70	78.62	55.12	64.68	37.50	61.71	46.65
Affiliation	84.21	82.97	83.52	74.09	70.42	72.16	78.24	70.04	73.86	42.68	81.38	55.99
Phone	89.78	92.58	91.10	74.86	83.08	78.72	77.91	81.67	79.71	55.79	72.63	63.11
Fax	92.51	89.35	90.83	73.03	57.49	64.28	77.18	54.99	64.17	84.62	79.28	81.86
Email	81.21	82.22	80.35	81.66	70.32	75.47	93.14	69.18	79.37	51.82	72.32	60.38
Address	87.94	84.86	86.34	77.66	72.88	75.15	86.29	69.62	77.04	55.68	76.96	64.62
Bsuniv	74.44	62.94	67.38	64.08	53.16	57.56	86.06	46.26	59.54	21.43	20.00	20.69
Bsmajor	73.20	58.83	64.20	67.78	53.68	59.18	85.57	47.99	60.75	53.85	18.42	27.45
Bsdate	62.26	47.31	53.49	50.77	34.58	40.59	68.64	18.23	28.49	17.95	16.67	17.28
Msuniv	66.51	51.78	57.55	59.81	40.06	47.49	89.38	34.77	49.78	15.00	8.82	11.11
Msmajor	69.29	59.03	63.35	69.91	56.56	61.92	86.47	49.21	62.10	45.45	20.00	27.78
Msdate	57.88	43.13	48.96	48.11	36.82	41.27	68.99	19.45	30.07	30.77	25.00	27.59
Phduniv	71.22	58.27	63.73	60.19	48.23	53.11	82.41	43.82	57.01	23.40	14.29	17.74
Phdmajor	77.55	62.47	67.92	71.13	51.52	59.30	91.97	44.29	59.67	68.57	42.11	52.17
Phddate	67.92	51.17	57.75	50.53	36.91	42.49	73.65	29.06	41.44	39.13	15.79	22.50
Overall	84.98	81.90	83.37	75.04	69.41	72.09	81.66	66.97	73.57	48.60	59.36	53.44

i.e., $P(Y|X)$ [87]. All components Y_i of Y are assumed to range over a finite label alphabet Y (as the properties defined in Figure 5.3). The conditional probability is formalized as:

$$P(y|x) = \frac{1}{Z(x)} \exp(\sum_{e,j} \lambda_j t_j(e, y|_e, x) + \sum_{v,k} \mu_k s_k(v, y|_v, x)), \tag{5.1}$$

where x is a data sequence, y is a label sequence, and $y|_e$ and $y|_v$ are the set of components of y associated with edge e and vertex v in the data sequence, respectively; t_j and s_k are feature functions; parameters λ_j and μ_k are coefficients respectively corresponding to the feature functions t_j and s_k, and are to be estimated from the training data; and $Z(x)$ is the normalization factor.

In tagging, a trained CRF model is used to find the sequence of tags Y^* having the highest likelihood $Y^* = \max_Y P(Y|X)$, with the Viterbi algorithm.

In training, the CRF model is built with labeled data and by means of an iterative algorithm based on Maximum Likelihood Estimation.

Feature Definition. Three types of features were defined: content features, pattern features, and term features. The features were defined for different kinds of tokens. Table 5.1 shows the defined features.

We can easily incorporate the defined features into the CRF model by defining Boolean-valued feature functions. Finally, 108,409 features were used in our experiments.

5.2.2   PROFILE EXTRACTION PERFORMANCE

For evaluating our unified profiling method, we randomly chose $1,000$ researcher names in total from our researcher network base. We used the method described in Section 5.2.1 to find the researchers' homepages or introducing pages. If the method cannot find a web page for a researcher, we remove the researcher name from the data set. We finally obtained 898 web pages. Seven human annotators conducted annotation on the web pages. A spec was created to guide the annotation process. For disagreements in the annotation, we conducted "majority voting." The annotated data set and the annotation specification are publicly available.[5]

In the experiments, we conducted evaluations in terms of precision, recall, and F1-measure (for definitions of the measures, see for example van Rijsbergen [153]).

We defined baselines for researcher profile extraction. We use the rule learning and the classification based approach as baselines. For the former approach, we employed the Amilcare system [28]. The system is based on a rule induction algorithm, called LP^2. For the later approach, we train a classifier for identifying the values of each property. We employed Support Vector Machines (SVM) [30] as the classification model.

To test how dependencies between different types of properties affect profiling, we also conducted experiments using the unified model by removing the transition features (Unified_NT).

Table 5.2 shows the five-fold cross-validation results. Our method clearly outperforms the baseline methods ($+29.93\%$ and $+9.80\%$, respectively, in terms of F1-score). We can also see that the performance of the unified method decreases (-11.28% by F1) when removing the transition features (Unified_NT).

We have implemented the proposed profiling approaches in ArnetMiner. By employing the proposed profiling approaches, we have extracted more than 30 million researcher profiles. Figure 5.4 shows an example researcher profile. We see that in the top of the profile page, some basic information (e.g., person photo, position, and affiliation) of the researcher has been correctly extracted from the homepage. Below is the research interest and evolution of the research interest discovered by our interest analysis approach. The right side of the page shows the social graph of the researcher based on his co-author relationships.

5.3   NAME DISAMBIGUATION

We integrated the publication data from existing online data sources, including DBLP bibliography,[6] ACM,[7] and others, covers more than 4 million papers from major computer science publication venues. In each data source, authors are identified by their names. For integrating the researcher profiles and the publications data, we use researcher names and the author names as the identifier. The process inevitably has the ambiguous problem.

---

[5]http://arnetminer.org/lab-datasets/profiling/
[6]http://dblp.uni-trier.de/
[7]dl.acm.org

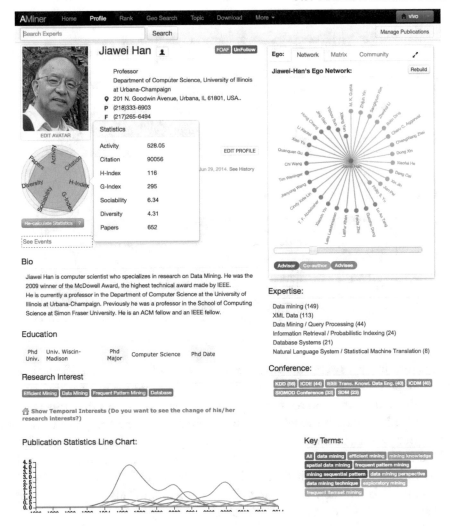

Figure 5.4: An example of researcher profile.

Here we give a formal definition of the name disambiguation task in our context. Given a person name a, we denote all publications having the author name a as $P = \{p_1, p_2, \quad, p_n\}$. For each publication p_i, it has six attributes as shown in Table 5.3.

For the authors of a paper $\{a_i^{(0)}, a_i^{(1)}, \quad, a_i^{(u)}\}$, we call the first author $a_i^{(0)}$ as the principal author and the others secondary authors. Suppose there exist k actual researchers $\{y_1, y_2, \quad, y_k\}$ having the name a, our task is then to assign these n publications to their real researcher y_h, where $h \in [1, k$.

Table 5.3: Attributes of each publication

Attribute	Description
$p_i.title$	title of p_i
$p_i.pubvenue$	published conference/journal of p_i
$p_i.year$	published year of p_i
$p_i.abstract$	abstract of p_i
$p_i.authors$	authors name set of p_i, $\{a_i^{(0)}, a_i^{(1)}, \cdots, a_i^{(u)}\}$
$p_i.references$	references of p_i

Table 5.4: Relationships between papers

R	W	Relation Name	Description
r_1	w_1	Co-Conference	$p_i.pubvenue = p_j.pubvenue$
r_2	w_2	Co-Author	$\exists r, s > 0, a_i^{(r)} = a_j^{(s)}$
r_3	w_3	Citation	p_i cites p_j or p_j cites p_i
r_4	w_4	Constraints	Feedbacks supplied by users
r_5	w_5	$\tau-$CoAuthor	$\tau-$extension co-authorship ($\tau > 1$)

We define five types of relationships between papers (Table 5.4). Relationship r_1 represents two papers are published at the same venue. Relationship r_2 means two papers have a secondary author with the same name, and relationship r_3 means one paper cites the other paper. Relationship r_4 indicates constraint-based relationships supplied via user feedbacks. For instance, the user can provide that two specific papers should be disambiguated to the same author. We use an example to explain relationship r_5. Suppose p_i has authors "David Mitchell" and "Andrew Mark," and p_j has authors "David Mitchell" and "Fernando Mulford." (We are going to disambiguate "David Mitchell.") If "Andrew Mark" and "Fernando Mulford" also co-author one paper, then we say p_i and p_j have a 2-co-author relationship.

Specifically, to test whether two papers have a $\tau-$co-author relationship, we construct a Boolean-valued matrix M, in which an element is 1 if its value is larger than 0; otherwise 0 (cf. Figure 5.5). In matrix M, $\{p_1, p_2, \cdots, p_n\}$ are publications with the principle author name a. $\{a_1, a_2, \cdots, a_p\}$ is the union set of all $p_i.authors \backslash a_i^{(0)}$, $i \in [1, n]$. Note that $\{a_1, a_2, \cdots, a_p\}$ does not include the principle author name $a_i^{(0)}$. Sub matrix M_p indicates the relationship between $\{p_1, p_2, \cdots, p_n\}$ and initially it is an identity matrix. In sub matrix M_{pa}, an element on row i and column j is equal to 1 if and only if $a_j \in p_i.authors$, otherwise 0. The matrix M_{ap} is symmetric to M_{pa}. Sub matrix M_a indicates the co-authorship among $\{a_1, a_2, \cdots, a_p\}$. The value on row i and column j in M_a is equal to 1 if and only if a_i and a_j coauthor one paper in our database (not limited in $\{p_1, p_2, \cdots, p_n\}$), otherwise 0. Then $\tau-$co-author can be defined easily based on $M^{(\tau+1)}$, where $M^{(\tau+1)} = M^{(\tau)} M$ with $\tau > 0$.

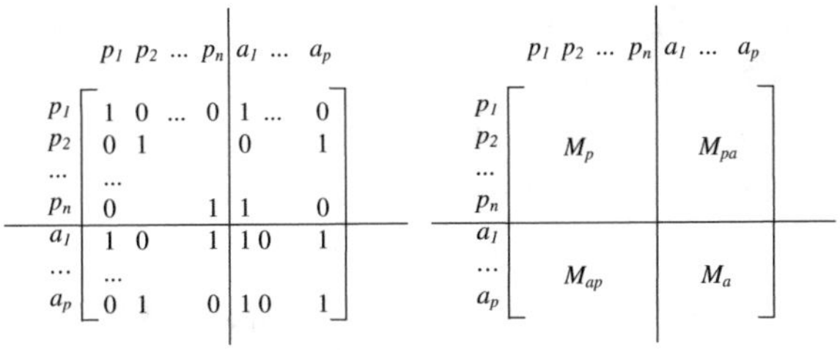

$$
\begin{array}{c}
\begin{array}{cccc|ccc}
p_1\ p_2\ \dots\ p_n & a_1\ \dots\ a_p
\end{array}\\
\begin{array}{c}
p_1\\p_2\\\dots\\p_n\\\hline a_1\\\dots\\a_p
\end{array}
\left[
\begin{array}{cccc|ccc}
1 & 0 & \dots & 0 & 1 & \dots & 0\\
0 & 1 & & & 0 & & 1\\
& \dots & & & & & \\
0 & & & 1 & 1 & & 0\\
\hline
1 & 0 & & 1 & 1 & 0 & 1\\
& \dots & & & & & \\
0 & 1 & & 0 & 1 & 0 & 1
\end{array}
\right]
\end{array}
\qquad
\begin{array}{c}
\begin{array}{cc|c}
p_1\ p_2\ \dots\ p_n & a_1\ \dots\ a_p
\end{array}\\
\begin{array}{c}
p_1\\p_2\\\dots\\p_n\\\hline a_1\\\dots\\a_p
\end{array}
\left[
\begin{array}{c|c}
M_p & M_{pa}\\
\hline
M_{ap} & M_a
\end{array}
\right]
\end{array}
$$

Figure 5.5: Matrix M for r_5 relationship.

Table 5.5: Abbreviate Name data set

Abbr. Name	# Publications	#Actual person	Abbr. Name	# Publications	#Actual Person
B.Liang	55	13	M.Hong	69	17
H. Xu	189	59	W. Yang	249	78
K. Zhang	320	40			

Table 5.6: Real name data set

Person Name	# Publications	#Actual Person	Person Name	# Publications	#Actual Person
Cheng Chang	12	3	Gang Wu	40	16
Wen Gao	286	4	Jing Zhang	54	25
Yi Li	42	21	Kuo Zhang	6	2
Jie Tang	21	2	Hui Fang	15	3
Bin Yu	66	12	Lei Wang	109	40
Rakesh Kumar	61	5	Michael Wagner	44	12
Bing Liu	130	11	Jim Smith	33	5

We empirically set the weight for each type of relationship. For example, we assign relationship r_4 with the highest weight. We assign co-author relationship r_2 a relatively high weight and w_5 as the τ power of w_2, i.e., $w_5 = w_2^\tau$. co-conference is a weak relationship, so we assign w_1 a small value. In our experiments, we set $w_1 \sim w_5$ as 0.2, 0.7, 0.3, 1.0, 0.7^τ, respectively.

The publication data with relationships can be modeled as a graph comprising of nodes and edges. Each attribute of a paper is attached to the corresponding node as a feature vector. For the vector, we use words (after stop words filtering and stemming) in the attributes of a paper as features and use occurring times as the values.

5.3.1   A UNIFIED PROBABILISTIC FRAMEWORK

We propose a probabilistic framework based on Hidden Markov Random Fields (HMRF) [11], which can capture dependencies between observations (here each paper is viewed as an observation). The disambiguation problem is cast as assigning a tag to each paper with each tag representing an actual researcher [140].

Specifically, we define the a-posteriori probability as the objective function. We aims at finding the maximum of the objective function. The five types of relationships are incorporated into the objective function. According to HMRF, the conditional distribution of the researcher labels y given the observation x (paper) has

$$P(y|x) = \frac{1}{Z(x)} \exp(-\sum_{i,h} D(x_i, y_h) - \sum_{i,j \neq i} (D(x_i, x_j) \sum_{r_k} w_k r_k(x_i, x_j))), \qquad (5.2)$$

where $D(x_i, y_h)$ is the distance between the paper x_i and the researcher y_h and $D(x_i, x_j)$ is the distance between papers x_i and x_j; $r_k(x_i, x_j)$ denotes a relationship between x_i and x_j; w_k is the weight of the relationship; and $Z(x)$ is the normalization factor.

The EM Algorithm. Three tasks are executed by the Expectation Maximization method: learning parameters of the distance measure, re-assignment of papers to researchers, and the update of researcher representatives y_h.

We define the distance function $D(x_i, x_j)$ as follows:

$$D(x_i, x_j) = 1 - \frac{x_i^T A x_j}{\|x_i\|_A \|x_j\|_A}, \text{where} \|x_i\|_A = \sqrt{x_i^T A x_j} \qquad (5.3)$$

(here A is a parameter matrix). For simplicity, we define it as a diagonal matrix.

The EM process can be summarized as follows: in the E-step, given the current researcher representatives, every paper is re-assigned to the researcher by maximize $P(Y|X)$. In the M-step, the researcher representative y_h is re-estimated from the assignments to maximize $P(Y|X)$ again, and the distance measure is updated to reduce the objective function.

In the initialization of our EM framework, we first cluster publications into disjoint groups based on the relationships between them, i.e., if two publications have a relationship, then they are assigned to the same researcher. Therefore, we first get λ groups. If λ is equal to our actual researcher number k, then these groups are used as our initial assignment. If $\lambda < k$, we choose $k - \lambda$ random assignments. If $\lambda > k$, we cluster the nearest group until there are only k groups left.

In the E-step, assignments of data points to researchers are updated to maximize the $P(Y|X)$. A greedy algorithm is used to sequentially update the assignment for each paper. The algorithm performs assignments in random order for all papers. Each paper x_i is assigned to y_h, $h \in [1, k]$ that minimize the function (equivalently to maximize $P(y_h|x_i)$):

$$f(y_h, x_i) = \sum_i D(x_i, y_h) + \sum_{i,j \neq i} (D(x_i, x_j) \sum_{r_k} w_k r_k(x_i, x_j)). \qquad (5.4)$$

The assignment of a paper is performed while keeping assignments of the other papers fixed. The assignment process is repeated after all papers are assigned. This process runs until no paper changes its assignment between two successive iterations.

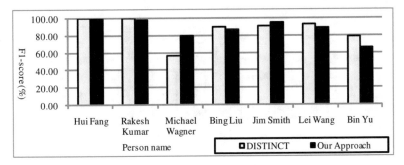

Figure 5.6: Comparison with existing method.

In the M-step, each researcher representative is updated by the arithmetic mean of its points:

$$y_h = \frac{\sum_{i:y_i=h} x_i}{\| \sum_{i:y_i=h} x_i \|_A}. \tag{5.5}$$

Then, each parameter a_{mm} in A is updated by (only parameters on the diagonal): $a_{mm} = a_{mm} + \eta \frac{\partial f(y_h, x_i)}{\partial a_{mm}}$, where:

$$\frac{\partial f(y_h, x_i)}{\partial a_{mm}} = -\sum_i \frac{\partial D(x_i, y_h)}{\partial a_{mm}} - \sum_{i,j \neq i} \left(\frac{\partial D(x_i, x_j)}{\partial a_{mm}} \sum_{r_k} w_k r_k(x_i, x_j) \right)$$

$$\frac{\partial D(x_i, x_j)}{\partial a_{mm}} = \frac{x_{im}x_{jm}\|x_i\|_A\|x_j\|_A - x_i^T A x_j \frac{x_{im}^2\|x_i\|_A^2 + x_{jm}^2\|x_j\|_A^2}{2\|x_i\|_A\|x_j\|_A}}{\|x_i\|_A^2\|x_j\|_A^2}.$$

5.3.2   NAME DISAMBIGUATION PERFORMANCE

Data Sets and Evaluation Measure. To evaluate our methods, we created two data sets from ArnetMiner, namely Abbreviate Name data set and Real Name data set. The first data set was collected by querying five abbreviated names in our database. All these abbreviated names are generated by simplifying the original names to its first name initial and last name. For example, "Cheng Chang" is simplified to "C. Chang." The simplification form is popular in bibliographic records. Statistics of this data set is shown in Table 5.5.

Another data set includes 14 real person names. In these names, some names only correspond to a few persons. For example, "Cheng Chang" corresponds to three actual persons and "Wen Gao" four; while some names seem to be popular. For example, there are 25 persons with the name "Jing Zhang" and 40 persons for "Lei Wang." Statistics of this data set are shown in Table 5.6.

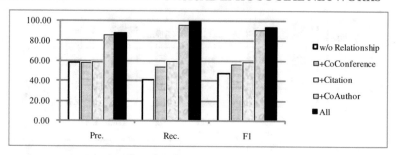

Figure 5.7: Contribution of relationships.

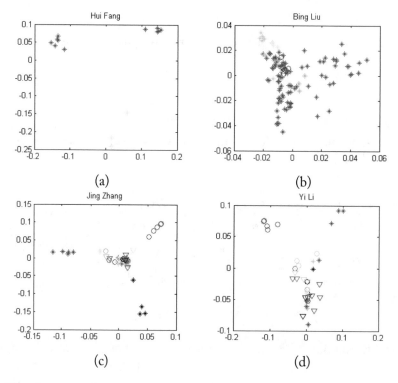

Figure 5.8: Distribution analysis.

Five human annotators conducted disambiguation on the papers. A spec was created to guide the annotation process. Each paper is labeled with a number indicating the actual person. The labeling work was carried out based on the publication lists on the authors' homepages, affiliations, and email addresses. For further disagreements in the annotation, we conducted "majority voting." The annotated data set can be downloaded here.[8]

---

[8]http://arnetminer.org/disambiguation

From the statistics we have found that the disambiguation results is extremely unbalance. For example, there are 286 papers authored by "Wen Gao" with 282 of them authored by Prof. Wen Gao from Institute of Computing at Chinese Academy of Science and only four papers are authored by the other three "Wen Gao."

We defined a baseline based on the hierarchical clustering algorithm. The baseline method is similar to that proposed by Tan et al. [139] except that Tan et al. [139] also utilize a search engine to help the disambiguation. We also compared our approach with the existing work, e.g., Yin et al. [171]. In all experiments, we suppose that the number of person k is provided empirically.

Results. We evaluated the performances of our method and the baseline methods on the two data sets. Table 5.7 shows the results. It can be seen that our method can significantly outperform the baseline method for name disambiguation (+4.54% on Abbr. Name data set and +10.75% on Real Name data set in terms of the average F1-score). More results were reported in Tang et al. [140].

Table 5.7: Results on name disambiguation (%)

Data Set	Person Name	Baseline			Our Approach		
		Prec.	Rec.	F1	Prec.	Rec.	F1
Abbr. Name	B. Liang	82.07	76.90	79.07	49.54	100.00	66.26
	H. Xu	65.87	59.48	71.27	32.77	100.00	49.37
	K. Zhang	75.67	60.27	67.84	71.03	100.00	83.06
	M. Hong	79.24	65.36	71.36	91.32	86.06	88.61
	W. Yang	71.30	62.83	66.99	52.48	99.86	68.81
Avg.		74.43	64.47	69.21	59.43	97.18	73.75
Real Name	Cheng Chang	100.00	100.00	100.00	100.0	100.0	100.0
	Wen Gao	96.60	62.64	76.00	99.29	98.59	98.94
	Yi Li	86.64	95.12	90.68	70.91	97.50	82.11
	Jie Tang	100.0	100.0	100.0	100.0	100.0	100.0
	Gang Wu	97.54	97.54	97.54	71.86	98.36	83.05
	Jing Zhang	85.00	69.86	76.69	83.91	100.0	91.25
	Kuo Zhang	100.0	100.0	100.0	100.0	100.0	100.0
	Hui Fang	100.0	100.0	100.0	100.0	100.0	100.0
	Bin Yu	67.22	50.25	57.51	86.53	53.00	65.74
	Lei Wang	68.45	41.12	51.38	88.64	89.06	88.85
	Rakesh Kumar	63.36	92.41	75.18	99.14	96.91	98.01
	Michael Wagner	18.35	60.26	28.13	85.19	76.16	80.42
	Bing Liu	84.88	43.16	57.22	88.25	86.49	87.36
	Jim Smith	92.43	86.80	89.53	95.81	93.56	94.67
Avg.		82.89	78.51	80.64	90.68	92.12	91.39

The baseline method suffers from two disadvantages: (1) it cannot take advantage of relationships between papers and (2) it relies on a fixed distance measure. Our framework benefits from the ability of modeling dependencies between assignment results.

We compared our approach with the approach DISTINCT proposed in Yin et al. [171]. We used the person names that were used both in Yin et al. [171] and in our experiments for comparisons. Figure 5.6 shows the comparison results. It can be seen that for some names, our approach significantly outperforms DISTINCT (e.g., "Michael Wagner"); while for some names our approach underperforms DISTINCT (e.g., "Bin Yu").

Feature Contribution Analysis. We further investigated the contribution of the defined relations for name disambiguation. We first evaluated the results of our approach by removing all relations. Then we added the relations: co-conference, citation, co-author, and $\tau-$co-author into our approach one by one. Figure 5.7 shows the results. "w/o Relationships" denotes our approach without any relationships. "+Co-Conference" denotes the results of by adding co-conference relationships. Likewise for the others. At each step, we observed improvements in terms of F1-score. We need note that without using relations the performances drop sharply (-15.65% on Abbr. Name and -44.72% on Real Name). This confirms us that a framework by integrating relationships for name disambiguation is needed and each defined relationships in our method is helpful.

We can also see that the co-author relationship makes major contributions ($+24.38\%$ by F1) to the improvements. co-conference and citation make limited contributions ($+0.68\%$ and $+0.61\%$) to the improvements on precision, but can obtain improvements ($+13.99\%$ and $+5.20\%$) on recall.

Distribution Analysis. Figure 5.8 shows several typical feature distributions in our data sets. The graphs were generated using a dimension reduction method described in Cai et al. [22]. The distributions can be typically categorized into: (1) papers of different persons are clearly separated ("Hui Fang," in Figure 5.8(a)). Name disambiguation on this kind of data can be solved pretty well by our approach and as well the baseline method; (2) publications are mixed together, however, there is a dominate author who writes most of the paper (e.g., "Bing Liu," in Figure 5.8(b)); our approach can achieve a F1-score of 87.36% however the baseline method results into a low accuracy (57.22% by F1); and (3) publications of different authors are mixed ("Jing Zhang" and "Yi Li," in Figure 5.8(c) and (d)). Our method can obtain 92.15% and 82.11% in terms of F1-measure; while the baseline method can only obtain a result 76.69% and 90.68% in terms of F1-measure, respectively.

Figure 5.9 shows a snapshot of the disambiguation result. The user searches for "Jie Tang" and the system returns three different persons on the top of the page and the background shows the detailed profile information of chosen person. The method runs in an offline mode and so far the system already generates the disambiguation results for more than 100,000 person names. Please note that this is an ongoing project. Visitors should expect the system to change.

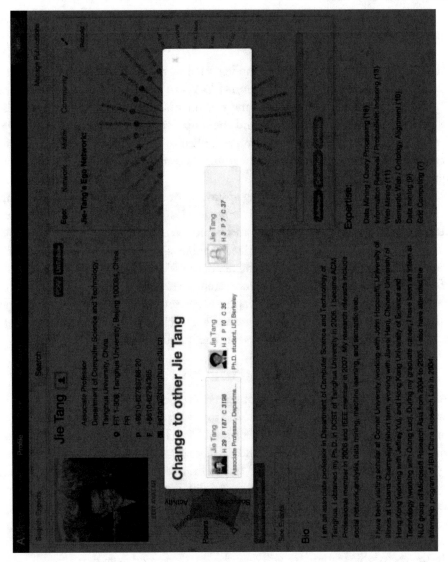

Figure 5.9: An example result of name disambiguation for "Jie Tang." There are three different "Jie Tang" automatically recognized by the proposed methods.

5.4   TOPIC MODELING

In academic search, representation of the content of text documents, authors interests, and conferences themes is a critical issue of any approach. Traditionally, documents are represented based on the "bag of words" (BOW) assumption. However, this representation cannot utilize the "semantic" dependencies between words. In addition, in academic search, there are different typed information sources, thus how to capture the dependencies between them becomes a challenging issue.

To deal with this problem, topic models such as probabilistic Latent Semantic Indexing (pLSI) [69] and Latent Dirichlet Allocation (LDA) [14] have been proposed. Both models can model the dependencies between words and documents. The Author-Topic model is proposed for modeling the content of documents and the interests of authors [122, 136]. However, all of the aforementioned models cannot be directly applied to the context of academic search, as they cannot capture all intrinsic dependencies in academic search such as dependencies between paper and conference.

We propose a unified topic modeling approach for simultaneously modeling characteristics of documents, authors, conferences, and dependencies between them. (For simplicity, we use conference to denote conference, journal, and book hereafter.)

The notations used here are summarized as follows. A document d is a vector of N_d words, $\mathbf{w}_d$, where each w_{di} is chosen from a vocabulary of size V, a vector of A_d authors $\mathbf{a}_d$, chosen from a set of authors of size A, and a published conference c_d. A collection of D documents is defined by $\mathbf{D} = \{(\mathbf{w}_1, \mathbf{a}_1, c_1), \cdots, (\mathbf{w}_D, \mathbf{a}_D, c_D)\}$. x_{di} indicates an author, chosen from $\mathbf{a}_d$, responsible for the ith word in document d. Here each author is associated with a distribution over topics Θ, chosen from a symmetric $Dirichlet(\alpha)$ prior. The number of topic is denoted as T.

5.4.1 OUR PROPOSED TOPIC MODELS

The proposed model is called Author-Conference-Topic (ACT) model. More specifically, different strategies can be employed to model the topic distributions (as shown in Figure 5.10) and thus the implemented models have different intuitions. In Figure 5.10(a), each author is associated with a mixture weights over topics and each word token in a paper and a conference stamp associated to each word token is generated from a sampled topic. In Figure 5.10(b), each author-conference pair is associated with a mixture weights over the topics and word tokens are then generated from the sampled topics. In Figure 5.10(c), each author is associated with topics and each word token is generated from a sampled topic, and then a conference is generated from the sampled topics of all word tokens in a paper.

In the rest of this section, we will describe the three models in detail.

Model One. In the first model, the conference information is viewed as a stamp for each word with the same value. The generative process in this ACT model (cf. Figure 5.10(a)) can be summarized as follows.

1. For each topic z, draw ϕ_z and ψ_z, respectively, from Dirichlet priors β_z and μ_z.

2. For each word w_{di} in document d:

   - draw an author x_{di} from $\mathbf{a}_d$ uniformly;

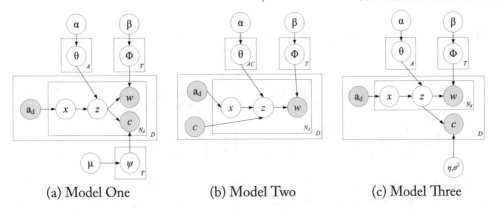

(a) Model One   (b) Model Two   (c) Model Three

Figure 5.10:   Graphical representation of the three Author-Conference-Topic (ACT) models.

- draw a topic z_{di} from a multinomial distribution $\theta_{x_{di}}$ specific to author x_{di}, where θ is generated from a Dirichlet prior α;
- draw a word w_{di} from multinomial $\phi_{z_{di}}$; and
- draw a conference stamp c_{di} from multinomial $\psi_{z_{di}}$.

In the generative process, all conference stamps in a document are observed as the same as the published venue of the document. In this way, the posterior distribution of topics depends on three modalities: authors, words, and conferences. Parameterizations in the ACT model are:

$$
\begin{aligned}
x|\mathbf{a}_d &\sim Uniform(\mathbf{a}_d) \\
\theta_{x_{di}}|\alpha &\sim Dirichlet(\alpha) \\
\phi_z|\beta &\sim Dirichlet(\beta) \\
\psi_z|\mu &\sim Dirichlet(\mu) \\
z_{di}|\theta_{x_{di}} &\sim Multinomial(\theta_{x_{di}}) \\
w_{di}|\phi_{z_{di}} &\sim Multinomial(\phi_{z_{di}}) \\
c_{di}|\psi_{z_{di}} &\sim Multinomial(\psi_{z_{di}}).
\end{aligned}
$$

Hence, in this ACT model, the joint probability of words $\mathbf{w}$, conferences $\mathbf{c}$, a set of corresponding latent topics $\mathbf{z}$, and an author mixture $\mathbf{x}$ is defined as

$$
P(\mathbf{x}, \mathbf{z}, \mathbf{w}, \mathbf{c}|\Theta, \Phi, \Psi, \mathbf{a}) = \prod_{d=1}^{D}\prod_{i=1}^{N_d}\frac{1}{A_d} \times \prod_{x=1}^{A}\prod_{z=1}^{T}(\theta_{xz}^{m_{xz}}\prod_{v=1}^{V}\phi_{zv}^{n_{zv}}\prod_{c=1}^{C}\psi_{zc_d}^{n_{zc_d}}), \quad (5.6)
$$

where m_{xz} is the number of times that topic z has been used associated with the chosen author x, n_{zv} is the number of times that word w_v has been generated by topic z, and n_{zc_d} is the number of times that conference c_d has been generated by topic z.

By placing a Dirichlet prior over Θ and another two over Φ and Ψ, and combining them into Equation (5.6) with further integrating over Θ, Φ and Ψ, we obtain:

$$P(\mathbf{x}, \mathbf{z}, \mathbf{w}, \mathbf{c}|\alpha, \beta, \mu, \mathbf{a}) = \prod_{d=1}^{D} \prod_{i=1}^{N_d} \frac{1}{A_d} \times \prod_{x=1}^{A} \frac{\Gamma(\sum_z \alpha_z)}{\prod_z \Gamma(\alpha_z)} \frac{\prod_z \Gamma(m_{xz} + \alpha_z)}{\Gamma(\sum_z (m_{xz} + \alpha_z))}$$

$$\times \prod_{z=1}^{T} \frac{\Gamma(\sum_v \beta_v)}{\prod_v \Gamma(\beta_v)} \frac{\prod_v \Gamma(n_{zv} + \beta_v)}{\Gamma(\sum_v (n_{zv} + \beta_v))} \times \prod_{z=1}^{T} \frac{\Gamma(\sum_c \mu_c)}{\prod_c \Gamma(\mu_c)} \frac{\prod_c \Gamma(n_{zc} + \mu_c)}{\Gamma(\sum_c (n_{zc} + \mu_c))}. \qquad (5.7)$$

It is intractable to directly compute Equation (5.7). A variety of algorithms have been proposed to conduct approximate inference, for example variational EM methods [14], Gibbs sampling [57, 136], and expectation propagation [57, 109]. We chose Gibbs sampling for its easy of implementation.

As for the hyperparameters α, β, and μ, one could estimate the optimal values by using a Gibbs EM algorithm [3, 108] or a variational EM method [14]. For some applications, topic models are sensitive to the hyperparameters and it is necessary to get the right values for the hyperparameters. In the applications discussed in this book, we found that the estimated topic models are not very sensitive to the hyperparameters. Thus, for simplicity, we took a fixed value (i.e., $\alpha = 50/T$, $\beta = 0.01$, and $\mu = 0.1$).

In the Gibbs sampling procedure, we need to calculate the posterior probability $P(z_{di}, x_{di}|\mathbf{z}_{-di}, \mathbf{x}_{-di}, \mathbf{w}, \mathbf{c}, \alpha, \beta, \mu)$ for the chosen author for sampling the topic for each word token. We begin with the joint probability of a data set (cf. Equation (5.7)), and using the chain rule, we can obtain the conditional probability as

$$P(z_{di}, x_{di}|\mathbf{z}_{-di}, \mathbf{x}_{-di}, \mathbf{w}, \mathbf{c}, \alpha, \beta, \mu) \propto \frac{m_{x_{di}z_{di}}^{-di} + \alpha_{z_{di}}}{\sum_z (m_{x_{di}z}^{-di} + \alpha_z)} \frac{n_{z_{di}w_{di}}^{-di} + \beta_{w_{di}}}{\sum_v (n_{z_{di}v}^{-di} + \beta_v)} \frac{n_{z_{di}c_d}^{-d} + \mu_{c_d}}{\sum_c (n_{z_{di}c}^{-d} + \mu_c)},$$
$$(5.8)$$

where the superscript $-t$ denote a quantity, excluding the current instance (the di-th word token or the conference stamp in the d-th document).

Given D documents, a set of topics $\mathbf{z}$, and hyperparameters α, β, and μ, the random variables ϕ (the probability of a word given a topic), ψ (the probability of a conference given a topic), and θ (the probability of a topic given an author) can be estimated via:

$$\phi_{zw_{di}} = \frac{n_{zw_{di}} + \beta_{w_{di}}}{\sum_v (n_{zv} + \beta_v)} \qquad (5.9)$$

$$\psi_{zc_d} = \frac{n_{zc_d} + \mu_{c_d}}{\sum_c (n_{zc} + \mu_c)} \qquad (5.10)$$

$$\theta_{xz} = \frac{m_{xz} + \alpha_z}{\sum_{z'} (m_{xz'} + \alpha_{z'})}. \qquad (5.11)$$

Model Two. The second model (cf. Figure 5.10(b)) is an alternative model to the first model. The difference is that it views the conference and authors in a document as a united information and thus each pair of author-conference is responsible for the sampled topics. The generative process of this model is:

1. for each topic z, draw ϕ_z from Dirichlet priors β_z; and

2. for each word w_{di} in document d:

   - draw an author-conference pair (x_{di}, c_d) from $\{\mathbf{a}_d, c_d\}$ uniformly;
   - draw a topic z_{di} from a multinomial distribution $\theta_{(x_{di}c_d)}$ specific to author-conference pair (x_{di}, c_d), where θ is generated from a Dirichlet prior α; and
   - draw a word w_{di} from multinomial $\phi_{z_{di}}$.

Parameterizations in this ACT model are similar as those in Model One, except that we chose a pair (x, c) from $\{\mathbf{a}_d, c\}$ uniformly and we will not sample the conference stamp in this model.

Thus, the joint probability of words $\mathbf{w}$, a set of corresponding latent topics $\mathbf{z}$, and an author mixture $\mathbf{x}$ is defined as

$$P(\mathbf{x}, \mathbf{z}, \mathbf{w} | \Theta, \Phi, \Psi, \mathbf{a}, \mathbf{c}) = \prod_{d=1}^{D} \prod_{i=1}^{N_d} \frac{1}{A_d * |c_d|} \times \prod_{xc=1}^{AC} \prod_{z=1}^{T} \prod_{v=1}^{V} \theta_{(xc)z}^{m_{(xc)z}} \phi_{zv}^{n_{zv}}, \tag{5.12}$$

where $A_d * |c_d|$ is the number of author-conference pairs for a document. (In practical, it is equal to A_d as a paper only has one publication venue.)

Again, by placing a Dirichlet prior over Θ and another one over Φ and putting them into Equation (5.12) by integrating out Θ and Φ, we can obtain:

$$P(\mathbf{x}, \mathbf{z}, \mathbf{w} | \alpha, \beta, \mathbf{a}, \mathbf{c}) = \prod_{d=1}^{D} \prod_{i=1}^{N_d} \frac{1}{A_d * |c_d|} \times \prod_{xc=1}^{AC} \frac{\Gamma(\sum_z \alpha_z)}{\prod_z \Gamma(\alpha_z)} \frac{\prod_z \Gamma(m_{(xc)z} + \alpha_z)}{\Gamma(\sum_z (m_{(xc)z} + \alpha_z))}$$

$$\times \prod_{z=1}^{T} \frac{\Gamma(\sum_v \beta_v)}{\prod_v \Gamma(\beta_v)} \frac{\prod_v \Gamma(n_{zv} + \beta_v)}{\Gamma(\sum_v (n_{zv} + \beta_v))}. \tag{5.13}$$

Similarly, we can calculate the posterior conditional probability $P(z_{di}, (xc)_{di} | \mathbf{z}_{-di}, \mathbf{x}_{-di}, \mathbf{c}_{-d}, \mathbf{w}, \alpha, \beta)$ using a Gibbs sampling procedure analogous to that in Model One

$$P(z_{di}, (xc)_{di} | \mathbf{z}_{-di}, \mathbf{x}_{-di}, \mathbf{c}_{-d}, \mathbf{w}, \alpha, \beta) \propto \frac{m_{(xc)_{di} z_{di}}^{-di} + \alpha_{z_{di}}}{\sum_z (m_{(xc)_{di} z}^{-di} + \alpha_z)} \frac{n_{z_{di} w_{di}}^{-di} + \beta_{w_{di}}}{\sum_v (n_{z_{di} v}^{-di} + \beta_v)}. \tag{5.14}$$

Model Three. Intuition of the third model (cf. Figure 5.10(c)) is derived from the observation that authors write a paper with a topic distribution and then they want to find a right conference to submit the paper. Thus, the corresponding generative process is:

1. for each topic z, draw ϕ_z from Dirichlet priors β_z;

2. for each word w_{di} in document d:

   - draw an author x_{di} from $\mathbf{a}_d$ uniformly;

   - draw a topic z_{di} from a multinomial distribution $\theta_{x_{di}}$ specific to author x_{di}, where θ is generated from a Dirichlet prior α; and

   - draw a word w_{di} from multinomial $\phi_{z_{di}}$.

3. draw a conference c_d from $z_{1:N_d}$ using a normal linear model $N(\eta^\top \tau, \sigma^2)$, where τ is a vector recording the normalized number of times of each topic sampled from document d. We define it as $\tau := (1/N_d) \sum_{i=1}^{N_d} |z_{di}|$.

In this model, the conference comes from a normal linear model. The covariates in this model are the empirical frequencies of the topics in the document. The regression coefficients on those frequencies constitute η. Note that we ignore the intercept term, which is used to guarantee adding a covariate equal to one, because in our model τ always sum to one. Thus, the difference of parameterizations of this model from Model One is that the conference stamp is sampled from a normal linear distribution after all topics were sampled for word tokens in a document.

Accordingly, the joint probability of words $\mathbf{w}$, conference stamps c, a set of corresponding latent topics $\mathbf{z}$, and an author mixture $\mathbf{x}$ is defined as

$$P(\mathbf{x}, \mathbf{z}, \mathbf{w}, \mathbf{c} | \Theta, \Phi, \eta, \sigma^2, \mathbf{a}) = \prod_{d=1}^{D} (P(c_d | z_{1:N_d}, \eta, \sigma^2)) \prod_{i=1}^{N_d} \frac{1}{A_d} \times \prod_{x=1}^{A} \prod_{z=1}^{T} \prod_{v=1}^{V} \theta_{xz}^{m_{xz}} \phi_{zv}^{n_{zv}}). \quad (5.15)$$

For computing the above equation using Gibbs sampling, there is a slight difference from that in Model One and Model Two, as we also need to estimate the values of η and σ^2. We use a Gibbs EM algorithm for this model.

In the E-step, for sampling the topic for each word token, the posterior probability $P(z_{di}, x_{di} | \mathbf{z}_{-di}, \mathbf{x}_{-di}, c_d, \mathbf{w}, \alpha, \beta)$ is calculated by

$$P(z_{di}, x_{di} | \mathbf{z}_{-di}, \mathbf{x}_{-di}, c_d, \mathbf{w}, \alpha, \beta, \eta, \sigma^2) \propto$$
$$P(c_d | z_{1:N_d}, \eta, \sigma^2) \frac{m_{x_{di} z_{di}}^{-di} + \alpha_{z_{di}}}{\sum_z (m_{x_{di} z}^{-di} + \alpha_z)} \frac{n_{z_{di} w_{di}}^{-di} + \beta_{w_{di}}}{\sum_v (n_{z_{di} v}^{-di} + \beta_v)}, \quad (5.16)$$

where

$$P(c_d|z_{1:N_d}, \eta, \sigma^2) = \frac{1}{\sqrt{2\pi\sigma^2}} e^{\left(-\frac{(c-\eta^\top\tau)^2}{2\sigma^2}\right)}. \tag{5.17}$$

In the M-step, given the sampled topics $\mathbf{z}$, the optimal η and σ^2 can be estimated by maximizing

$$argmx_{\eta,\sigma} \log P(\mathbf{x}, \mathbf{z}, \mathbf{w}, \mathbf{c}|\alpha, \beta, \eta, \sigma^2). \tag{5.18}$$

Specifically, η is updated by

$$\eta_{new} \leftarrow (E[A^\top A])^{-1} E[A]^\top \mathbf{c} \tag{5.19}$$

and σ^2 is updated by

$$\sigma_{new}^2 \leftarrow (1/D)\{\mathbf{c}^\top\mathbf{c} - \mathbf{c}^\top E[A](E[A^\top A])^{-1}E[A]^\top\mathbf{c}\}, \tag{5.20}$$

where $E[.]$ is the expectation of the variables; A is a matrix of $D \times T$ with the d-th row is $E[\tau] = \bar{\phi} := (1/N_d)\sum_{i=1}^{N_d} \phi_{di}$ and $E[A^\top A] = \sum_{d=1}^{D} E[\tau_d \tau_d^\top]$ is a $T \times T$ matrix, where $E[\tau_d \tau_d^\top]$ is defined as:

$$E[\tau_d \tau_d^\top] = (1/N_d^2)(\sum_{i=1}^{N_d}\sum_{j\neq i} \phi_{di}\phi_{dj}^\top + \sum_{i=1}^{N_d} \text{diag}\{\phi_{di}\}) \tag{5.21}$$

with $\text{diag}\{\phi_{di}\}$ indicating a matrix with diagonal as the vector of ϕ_{di}. Note that ϕ_{di} denotes a vector of probabilities of topics generating word w_{di}. (We omitted the details of derivation of the update Equations (5.19) and (5.20). Interested reader is referred to [13] and [108].)

We estimate the ACT models in an offline mode before applying it to the academic search. The first two proposed models result into a complexity of $O(MD\bar{N}_d T \bar{A}_d)$ with slight differences, where M is the number of sampling times, $\bar{N}_d$ is the average number of word tokens in a document, and $\bar{A}_d$ is the average number of authors. In most cases, the number $\bar{A}_d$ is negligible to the final complexity. The third model has to estimate the optimal values for η and σ^2, which leads to a higher complexity $O(MD\bar{N}_d T + MD^2 T^2)$. Note that $E[A]$, $E[A^\top A]$ can be computed in the previous E-step (while in the sampling process).

5.5   EXPERTISE SEARCH

In academic search, the goal is to find the expertise authors, expertise papers, and expertise conferences for a given query [147, 175]

Based on the proposed ACT models, we can calculate the likelihood of the a document model generating a word using the following equation (we use ACT Model 1 as the example):

$$P_{ACT1}(w|d,\theta,\phi) = \sum_{z=1}^{T} \sum_{x=1}^{A_d} P(w|z,\phi_z)P(z|x,\theta_x)P(x|d), \qquad (5.22)$$

where all probabilities on the right side of the equation are estimated in the ACT model. The likelihood of an author model and a conference model generating a word can be similarly defined.

However, as a topic in the LDA-style model represent a combination of words, it may not be as precise as representation as words in non-topic models like the language model. Therefore, only using ACT itself for modeling is too coarse for academic search [160]. Actually, our preliminary experiments also show that employing only the ACT or LDA models to information retrieval hurts the retrieval performance. Finally, we derive a combination form of the ACT models and the language model:

$$P(w|d) = P_{LM}(w|d) \times P_{ACT}(w|d), \qquad (5.23)$$

where $P_{LM}(w|d)$ is the generating probability of word w from document d by the language model. It is defined as [173]:

$$P(w|d) = \frac{N_d}{N_d + \lambda} \cdot \frac{tf(w,d)}{|d|} + (1 - \frac{N_d}{N_d + \lambda}) \cdot \frac{tf(w,\mathbf{D})}{|\mathbf{D}|}, \qquad (5.24)$$

where $|d|$ is the length of document d, $tf(w,d)$ is the word frequency (i.e., number of words) of word w in d, $|\mathbf{D}|$ is the number of word tokens in the whole collection, and $tf(w,\mathbf{D})$ is the word frequency of word w in the whole collection $\mathbf{D}$. λ is the Dirichlet prior and is common to set it according to the average document length in the document collection.

Finally, given a query q, $P(q|d)$ can be computed by $P(q|d) = \Pi_{w \in q} P(w|d)$. Similarly, we can define $P(q|a)$ for authors and $P(q|c)$ for conferences in a analogous way:

$$P(q|a) = P_{LM}(q|a) \times P_{ACT}(q|a) \qquad (5.25)$$

$$P(q|c) = P_{LM}(q|c) \times P_{ACT}(q|c) \qquad (5.26)$$

where a and c is represented by a collection of papers published by author a and on conference c, respectively.

5.5.1   DATA SETS AND EVALUATION MEASURES

We have collected a list of queries from the query log of ArnetMiner for evaluation purposes. Specifically, we selected the most frequent queries from the log of ArnetMiner (by removing overly specific or lengthy queries, e.g., "A Convergent Solution to Tensor Subspace Learning"). We also normalized similar queries (e.g., "Web Service" and "Web Services" to "Web Service").

We conducted our experiments on a subset of the data (including $14,134$ persons, $10,716$ papers, and $1,434$ conferences) from ArnetMiner. For evaluation, it is difficult to find a standard data set with ground truth. As a result, we use the method of pooled relevance judgments [20] together with human judgments. Specifically, for each query, we first pooled the top 30 results from three similar (academic search) systems (Libra, Rexa, and ArnetMiner) into a single list. Then, two faculties and five graduate students from the authors' lab provided human judgments. Four grade scores (3, 2, 1, and 0) were assigned, respectively, representing definite expertise, expertise, marginal expertise, and not expertise. For example, for annotating persons, assessments were carried out mainly in terms of how many publications she/he has published related to the given query, how many top conference papers he or she has published, what distinguished awards she/he has been awarded. Finally, the judgment scores were averaged to construct the final truth ground. The data set is available online.

In all experiments, we conducted evaluation in terms of P@5, P@10, P@20, R-pre, and mean average precision (MAP). Readers are referred to Buckley and Voorhees [20] and Craswell et al. [33] for details of the measures.

We use language model (LM) [173], LDA [14], and Author-Topic model [122, 136] as the baseline methods. For language model, we use Equation (5.24) to calculate the relevance between a query term and a document and for LDA, we use a similar equation with Equation (5.23) to calculate the relevance of a term and a document. AT model can simultaneously capture dependencies between words, documents, and authors. We use similar equations with Equation (5.23) and (5.25) to calculate the relevance of a query term with a document and an author respectively.

For the LDA and AT models, we performed model estimation on the same data set as that for the ACT models. We empirically set the number of topics as 80 for LDA, AT, and ACT models by minimizing the perplexity [7], a standard measure for estimating the performance of a probabilistic model (the lower the best), of the estimated topic models. One can also use some solution like [152] to automatically estimate the number of topics.

All experiments were carried out on a Server running Windows 2003 with two Dual-Core Intel Xeon processors (3.0 GHz) and 2GB memory. It needs 20 and 50 min, respectively, estimating the LDA model and AT model, 65 min for estimating ACT Model 1, 1.2 hr for ACT Model 2, and about 3 hr for ACT Model 3 (for 2000 sampling iterations).

5.5.2 RESULTS

Expertise Search

We evaluated performances of our proposed methods (referred to as ACT1, ACT2, and ACT3) and the baseline methods (LM, LDA, and AT) using the collected evaluation queries.

Table 5.8 shows the experimental results of retrieving papers, authors, and conferences using our proposed methods and the baseline methods. We see that all of our proposed three methods outperform the baseline methods (LM, LDA, and AT). For academic search, the improvements over the language model based method range from +5.1% to +9.8% in terms of MAP and improvements over the pLSI based method range from +0.4% to +5.1% in terms of MAP. Based on all the other evaluation measures, our method consistently performs better than

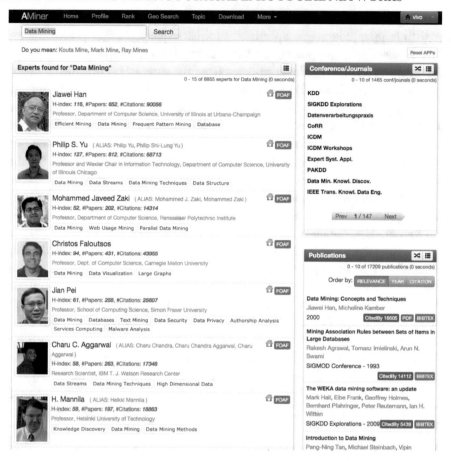

Figure 5.11: An example result of expert finding for "Data Mining." The left side lists experts ranked by the proposed method, the top-right lists ranked publication venues, and the bottom-right lists ranked papers.

the three baseline methods. LDA only models dependencies between documents and words and thus can support only paper search; while AT only models dependencies between documents, words, and authors, and thus support paper search and author search. Both of the two methods underperform our proposed unified models. Our models benefit from the ability of capturing all kinds of dependencies among words, papers, authors, and conferences.

We can also see that ACT1 outperforms ACT2 and ACT3. ACT1 achieves the best performance in terms of P@5 (+10.8%, +9.2%, +5.0%, and +4.8% against LM, pLSI, ACT2, and ACT3, respectively). ATC2 suffers from a sparse problem when sampling a pair of author-conference from a document. In our experiment, the possible number of pairs scales up to more than 10 million, which makes the associated topics to each pair is very sparse. Thus, the estimated topics

Table 5.8: Performances of five expert finding approaches (%)

Method	Object	P@5	P@10	P@20	R-pre	MAP
LM	Paper	37.5	33.8	32.5	8.7	42.1
	Author	55.0	37.5	21.3	49.0	62.8
	Conference	47.5	30	19.4	40.8	51.0
	Average	46.7	33.8	24.4	32.8	52.0
pLSI	Paper	32.5	33.8	30	9.7	40.4
	Author	65.0	40.0	22.5	60.4	75.5
	Conference	47.5	36.3	21.3	45.1	54.1
	Average	48.3	36.7	24.6	38.4	56.7
ACT1	Paper	37.5	40.0	36.9	13.7	43.8
	Author	80.0	43.8	23.1	70.0	78.4
	Conference	55.0	36.3	20.6	53.1	63.3
	Average	57.5	40.0	26.9	45.6	**61.8**
ACT2	Paper	37.5	28.8	28.1	9.1	41.6
	Author	65.0	38.8	23.8	56.6	66.8
	Conference	55.0	33.8	20.6	51.6	62.9
	Average	52.5	33.8	24.2	39.1	57.1
ACT3	Paper	35.0	37.5	36.3	10.6	43.1
	Author	67.5	41.3	24.4	60.7	69.5
	Conference	52.5	36.3	20.6	53.1	61.2
	Average	52.7	38.3	27.1	41.5	57.9

cannot accurately characterize the author. As a result, for author search, ACT2 underperforms ACT1 by −11.6% in terms of MAP. ACT3 underperforms ACT1 by P@5 and P@10 but obtains the best performance in terms of P@20 and P@30.

For comparison purposes, we also evaluated the results returned by two similar systems: Libra.msra.cn and Rexa.info. The average MAP obtained by Libra and Rexa on our collected queries are respectively 51.0% and 46.2%. We see that our methods clearly outperform the two systems.

Tables 5.9 and 5.10 show the example of retrieved results for the query "Information Extraction" and "Semantic Web." We can see some interesting patterns. For example, we have found that in the topic found in Table 5.11, "ISWC" has the highest generative probability from the topic "Semantic Web" but for the query "Semantic Web," it is retrieved in the second place while "WWW" is returned as the top one. This is because many experts on "Semantic Web" also has published many papers on "WWW." The combination of the ACT model and the language model re-rank the results of ACT model.

Figure 5.11 shows an example of expert finding in ArnetMiner. The user tries to find experts on "Data Mining." The system returns experts ranked by the proposed ranking method. In

Table 5.9:   Retrieved results of "Information Extraction" by ACT1

Person	Conference	Paper
Raymond J. Mooney	AAAI	Multistrategy Learning for Information Extraction
Nicholas Kushmerick	IJCAI	Adaptive Information Extraction: Core Technologies for Information Agents
Andrew McCallum	ICML	Integrating Information to Bootstrap Information Extraction from Web Sites
Dayne Freitag	ACL	Automatically Constructing a Dictionary for Information Extraction Tasks
Ellen Riloff	SIGIR	Machine Learning for Information Extraction in Informal Domains
Dan I. Moldovan	EACL	Collective Information Extraction with Relational Markov Networks
Fabio Ciravegna	COLING	Exploiting Subjectivity Classification to Improve Information Extraction
Alexiei Dingli	IIWeb	Evaluating machine learning for information extraction
Daniel S. Weld	Machine Learning	Multi-level Boundary Classification for Information Extraction
Bernd Thomas	ECML	Adaptive information extraction for document annotation in amilcare

Table 5.10:   Retrieved results of "Semantic Web" by ACT1

Person	Conference	Paper
Ian Horrocks	WWW	Metamodeling Architecture of Web Ontology Languages
Steffen Staab	ISWC	Three theses of representation in the semantic web
Stefan Decker	Description Logics	Learning Ontologies for the Semantic Web
York Sure	IEEE Intelligent Systems	DAML+OIL: a Description Logic for the Semantic Web
Carole A. Goble	ESWC	Querying the Semantic Web: A Formal Approach
Rudi Studer	J. Web Sem.	A software framework for matchmaking based on semantic web technology
Peter F. Patel-Schneider	ESWS	The Networked Semantic Desktop
Jeff Z. Pan	AAAI	OilEd: a Reason-able Ontology Editor for the Semantic Web
Sergio Tessaris	IJCAI	A Scalable Framework for the Interoperation of Information Sources
Sean Bechhofer	IEEE Data Eng. Bull.	OntoEdit: Collaborative Ontology Development for the Semantic Web

addition, the top-right of the screen lists ranked publication venues, and the bottom-right lists ranked papers.

Association Search

Given a social network, the association is defined as as follows.

Association: Given a social network $G = (V, E)$, an association $\alpha(a_i, a_j)$, $a_i, a_j \in V$, is a sequence of relations $\{e_{i1}^r, e_{12}^r, \cdots, e_{lj}^r\}$ satisfying $e_{m(m+1)}^r \in E$ for $m = 1, 2, \cdots, l - 1$, where a_i and a_j are the source and the target persons, respectively.

We assume that no person will appear on a given association more than one time. Each association is assigned with a score representing its goodness.

Association Search: Given an association query (a_i, a_j), association search is to find possible associations $\{\alpha_k(a_i, a_j)\}$ from a_i to a_j and rank the associations according to their goodness.

From the definition, we see that there are two subtasks in association search: finding associations between the two persons and ranking the associations. Given a large-scale social network, to find all possible associations is obviously an NP-hard problem. In this paper, we focus on finding the most "goodness" associations. Hence, the problem becomes how to evaluate the goodness of an association and one key issue is how to calculate the distance between two persons. Based on the modeling results (cf. Section 5.4), we define the distance between authors a_i and a_j as the

Table 5.11: Four topics discovered by ACT models on the ArnetMiner publication data set. Each topic is shown with with the top 10 words and their corresponding conditional probabilities. Below are top 6 authors and top 6 conferences associated with each topic. The titles are our interpretation of the topics. (CL - Computational Linguistics, JMLR - Journal of Machine Learning Research, MLSS - Machine Learning Summer School, JAIR - J. Artif. Intell. Res., and LNLP - Learning for Natural Language Processing.) (Continues.)

Topic #5 (Model 1) "Natural language processing"		Topic #10 (Model 1) "Semantic web"		Topic #16 (Model 1) "Machine learning"		Topic #24 (Model 1) "Information extraction"	
language	0.034820	semantic	0.068226	learning	0.058056	learning	0.065259
parsing	0.023766	web	0.048847	classification	0.018517	information	0.043527
natural	0.019029	ontology	0.043160	boosting	0.015881	extraction	0.033592
learning	0.015871	knowledge	0.041497	machine	0.017797	web	0.019311
approach	0.012712	learning	0.013431	feature	0.013904	semantic	0.011860
grammars	0.012712	framework	0.012095	classifiers	0.013904	text	0.010618
processing	0.011923	approach	0.011427	margin	0.013245	rules	0.010618
text	0.011923	based	0.010758	selection	0.012586	relational	0.009376
machine	0.011133	management	0.010090	algorithm	0.025086	logic	0.009376
probabilistic	0.010343	reasoning	0.009502	kernels	0.011269	programming	0.008755
Yuji Matsumoto	0.001389	Steffen Staab	0.005863	Robert E. Schapire	0.004033	Raymond J. Mooney	0.010346
Eugene Charniak	0.001323	Enrico Motta	0.004365	Yoram Singer	0.003318	Andrew McCallum	0.004074
Rens Bod	0.001323	York Sure	0.003713	Thomas G. Dietterich	0.002472	Craig A. Knoblock	0.003492
Brian Roark	0.001190	Nenad Stojanovic	0.001824	Bernhard Scholkopf	0.001496	Nicholas Kushmerick	0.002457
Suzanne Stevenson	0.001124	Alexander Maedche	0.001824	Alexander J. Smola	0.001301	Ellen Riloff	0.002199
Anoop Sarkar	0.001058	Asuncion Gomez-Perez	0.001694	Ralf Schoknecht	0.001236	William W. Cohen	0.002134
Claire Cardie	0.000992	Frank van Harmelen	0.001563	Michael I. Jordan	0.001106	Dan Roth	0.001487
ACL	0.253487	ISWC	0.125291	NIPS	0.289761	AAAI	0.295846
COLING	0.234435	EKAW	0.122379	JMLR	0.206583	IJCAI	0.192995
CL	0.118136	IEEE Intelligent Systems	0.071418	ICML	0.156389	ICML	0.060567
ANLP	0.060423	CoopIS/DOA/ODBASE	0.065594	COLT	0.096157	KDD	0.058551
CoRR	0.058674	K-CAP	0.054674	Neural Computation	0.023017	JAIR	0.046451
COLING-ACL	0.036814	ESWS	0.023369	MLSS	0.011545	ECML	0.033006
NAACL	0.035065	WWW	0.016817	Machine Learning	0.010827	IIWeb	0.026956

Table 5.11: *(Continued.)* Four topics discovered by ACT models on the ArnetMiner publication data set. Each topic is shown with the top 10 words and their corresponding conditional probabilities. Below are top 6 authors and top 6 conferences associated with each topic. The titles are our interpretation of the topics. (CL - Computational Linguistics, JMLR - Journal of Machine Learning Research, MLSS - Machine Learning Summer School, JAIR - J. Artif. Intell. Res., and LNLP - Learning for Natural Language Processing.) *(Continues.)*

Topic #52 (Model 2)		Topic #2 (Model 2)		Topic #30 (Model 2)		Topic #75 (Model 2)	
language	0.209963	web	0.156658	learning	0.212239	information	0.265864
natural	0.074298	semantic	0.080464	reinforcement	0.082713	extraction	0.152239
processing	0.053100	ontology	0.053723	hierarchical	0.037838	induction	0.014540
machine	0.017064	service	0.037686	machine	0.024579	ranking	0.011651
methodology	0.012825	automatic	0.028723	robocup	0.015400	document	0.010688
management	0.012825	applications	0.015788	discovery	0.013361	agent	0.009726
boolean	0.008585	representation	0.014016	large	0.009281	incremental	0.007800
base	0.007525	query	0.013307	survey	0.007241	probabilistic	0.006837
learning	0.005405	database	0.011890	multiple	0.007241	boosting	0.005874
computer	0.004346	context	0.011535	research	0.006221	incomplete	0.005874
Raymond J. Mooney	0.001013	Ian Horrocks	0.000421	Thomas G. Dietterich	0.001743	Ellen Riloff	0.001107
Eric Brill	0.000802	Steffen Staab	0.000408	Peter Stone	0.001428	Fabio Ciravegna	0.000794
Rolf Schwitter	0.000697	Raphael Volz	0.000303	Andrew G. Barto	0.000904	Nicholas Kushmerick	0.000690
Michael Collins	0.000592	Dieter Fensel	0.000290	Sridhar Mahadevan	0.000694	Stephen Soderland	0.000690
Claire Cardie	0.000592	Nicholas Kushmerick	0.000276	Raymond J. Mooney	0.000694	Andrew McCallum	0.000586
Christopher D. Manning	0.000592	William W. Cohen	0.000263	Ralf Schoknecht	0.000694	Dayne Freitag	0.000586
Cynthia A. Thompson	0.000487	Carole A. Goble	0.000250	Manuela M. Veloso	0.000485	Wai Lam	0.000586
ACL	0.028150	ISWC	0.028819	ICML	0.085238	AAAI	0.035059
CL	0.018401	WWW	0.022842	Machine Learning	0.036264	ACL	0.021953
AAAI	0.018401	ACL	0.020849	AAAI	0.030434	ICML	0.020861
CoRR	0.017183	AAAI	0.020849	NIPS	0.029268	IJCAI	0.020861
CICLing	0.017183	IJCAI	0.019219	RoboCup	0.021105	SIGIR	0.020861
ICML	0.014745	ICML	0.014509	ECML	0.016441	CIKM	0.015400
Machine Learning	0.014745	Web Intelligence	0.013060	IJCAI	0.016441	Machine Learning	0.014308

Table 5.11: *(Continued.)* Four topics discovered by ACT models on the ArnetMiner publication data set. Each topic is shown with the top 10 words and their corresponding conditional probabilities. Below are top 6 authors and top 6 conferences associated with each topic. The titles are our interpretation of the topics. (CL - Computational Linguistics, JMLR - Journal of Machine Learning Research, MLSS - Machine Learning Summer School, JAIR - J. Artif. Intell. Res., and LNLP - Learning for Natural Language Processing.)

Topic #20 (Model 3)		Topic #34 (Model 3)		Topic #15 (Model 3)		Topic #58 (Model 3)	
language	0.209132	web	0.282681	learning	0.388622	information	0.372220
processing	0.090411	semantic	0.251191	machine	0.082680	extraction	0.200865
natural	0.081279	services	0.067097	induction	0.045596	rules	0.009606
conceptual	0.013293	documents	0.005733	theories	0.009355	forms	0.004414
role	0.013293	search	0.004925	tasks	0.008512	answer	0.004414
syntactic	0.011263	collaborative	0.004118	scientific	0.008512	hybrid	0.003548
information	0.010249	incremental	0.004118	noise	0.006827	execution	0.003548
computer	0.009234	soccer	0.004118	empirical	0.006827	oriented	0.003548
identifying	0.009234	actions	0.004118	quantitative	0.005984	general	0.003548
multiple	0.008219	named	0.004118	introduction	0.005984	discriminative	0.003548
Eric Brill	0.001847	Steffen Staab	0.002003	Pat Langley	0.003963	Fabio Ciravegna	0.002432
Claire Cardie	0.000904	Ian Horrocks	0.002003	Raymond J. Mooney	0.003450	Dayne Freitag	0.001608
Christopher D. Manning	0.000904	Carole A. Goble	0.001493	Russell Greiner	0.001295	Craig A. Knoblock	0.001505
Hugo Liu	0.000799	Katia P. Sycara	0.001391	Nicholas Kushmerick	0.001295	Ellen Riloff	0.001403
Andres Montoyo	0.000799	Enrico Motta	0.001187	Manuela M. Veloso	0.000988	Nicholas Kushmerick	0.001300
Raymond J. Mooney	0.000799	Raphael Volz	0.001084	Larry A. Rendell	0.000988	Raymond J. Mooney	0.001197
Rens Bod	0.000694	Nicholas Kushmerick	0.000982	Xindong Wu	0.000988	Stephen Soderland	0.001094
ACL	0.046487	ISWC	0.055665	ICML	0.057446	ACL	0.034960
CL	0.022142	WWW	0.024292	Machine Learning	0.055566	AAAI	0.034960
CoRR	0.020983	ACL	0.022499	AAAI	0.040523	IJCAI	0.023339
ICML	0.019824	IEEE Intelligent Systems	0.021603	IJCAI	0.031121	CIKM	0.017529
CICLing	0.018665	AAAI	0.018017	JMLR	0.020778	SIGIR	0.017529
NIPS	0.015187	RuleML	0.015328	ECML	0.016077	ICML	0.016560
COLING	0.014027	Web Intelligence	0.015328	ACL	0.014197	KDD	0.015592

symmetric KL divergence between the topics distribution conditioned on each of the authors:

$$sKL(a_i, a_j) = \sum_{z=1}^{T} (\theta_{a_i z} log \frac{\theta_{a_i z}}{\theta_{a_j z}} + \theta_{a_j z} log \frac{\theta_{a_j z}}{\theta_{a_i z}}), \qquad (5.27)$$

where θ_{az} is calculated by Equation (5.11).

We use the accumulated value of distances between authors on an association path as the score of the association. We call the association with the smallest score as the shortest association and our goal is to find the near-shortest associations. By near-shortest associations, we mean associations whose scores are within a factor of $(1 + \gamma)$ of the score of the shortest association for some user-defined $\gamma > 0$.

We formalize the association search problem as that of near-shortest associations and employ a two stage approach to find the near-shortest associations.

(1) Shortest association finding. It aims at finding shortest associations from all persons $a \in V \backslash a_j$ in the network to the target person a_j (the score of the shortest association from a_i to a_j is denoted as L_{min}). In a graph, the shortest path between two nodes can be found using the state-of-the-art algorithms, for example, Dijkstra algorithm. However, we are dealing with a large-scale social network, the conventional Dijkstra algorithm results in a high time complexity of $O(n^2)$. We use a heap-based Dijkstra algorithm to quickly find the shortest associations that can achieve a complexity of $O(nlogn)$.

(2) Near-shortest associations finding. Based on the shortest association score $L_{min} > 0$ found in Step 1 and a pre-defined parameter γ, the algorithm requires enumeration of all associations that are less than $(1 + \gamma)L_{min}$ by a depth-first search. We constrain the length of an association to be less than a pre-defined threshold. This length restriction can reduce the computational cost, and each association score is calculated by summing the distances (computed by Equation (5.27)) between authors on the association. Finally, all associations are ranked based on the scores according to the policy of "the lower the best."

Table 5.12: Performances of association search (Second)

Test Set	Method	Total Time	Avg. Time
	Brute Force	483,708	483.71
Average	Two-stage Baseline	1,161,441	1161.44
	Our Approach	2,941	2.94

Association Search Performance. As there is no a standard data set for evaluating the association search. Here we focus on evaluating the effectiveness of the proposed approach. We created 9 test sets, each of them containing 369–1, 000 association queries (i.e., person pairs). An association query is composed of a source person and a target person.

The test sets were created as follows. We randomly selected 1, 000 person pairs from the researcher network and created the first test set. For the other eight test sets, we collected person

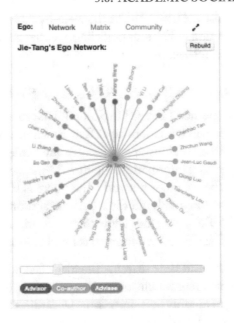

Figure 5.12: An example result of advisor-advisee relationship mining.

lists of different research areas (e.g., Data Mining (DM) and Semantic Web (SW)) and selected person pairs from different research areas.

We use the average running time on a test set to evaluate our approach and to compare with other methods. Running times do not include the time required to load the social network and output the associations.

We used the method of brute force enumeration as the baseline. In this method, we directly conducted depth-first search on the social network to find associations with length less than the threshold *max_length*. We also defined a two-stage method as the baseline. In the first stage of the method, we make use of the conventional Dijkstra algorithm [40] to find shortest paths and in the second stage we use depth-first search to find associations with length less than the threshold *max_length*.

Table 5.12 shows the results on the test sets. We see that our approach achieve high performance in all of the association search tasks. In terms of the average time, our approach can find associations in less than 3 s.

5.6   ACADEMIC SOCIAL NETWORK MINING

We now in particular discuss how the introduced technologies in Chapters 2–4 are applied in ArnetMiner. For social tie analysis, we applied the proposed unsupervised method TPFG (cf. Section 2.3.2) to automatically infer advisor-advisee relationships. For social influence, we use several case studies to demonstrate how effective the topic-based social influence graphs learned by the method presented in Section 3. For user modeling, we use the method described in Section 4 to extract authors' research interests.

5.6.1   MINING ADVISOR-ADVISEE RELATIONSHIPS

We applied the introduced method (TPFG) in Section 2.3.2 to analyze the roles of authors and discover the likely advisor-advisee relationships. Experimental results show that the proposed approach infer advisor-advisee relationships efficiently and achieves a state-of-the-art accuracy (80-90%) without any supervised information. Figure 5.12 shows the results of inferring advisor-advisee relationships from the coauthor network. The figure is the ego-network of the centered author "Jie Tang." The red colored link indicates that the author on the other end of the link is the identified advisor of "Jie Tang" and the yellow colored relationships indicate those authors on the other end are advisee of "Jie Tang," while green colored links indicate colleague relationships.

Applications: Expert finding and Bole search. The identified advisor-advisee relationships can help with many applications. Here we illustrate one application on bole search [169], a specific expert finding task, aiming to identify best supervisors (according to their nurture ability [110]) in a specific research field. The task requires advisor-advisee relationships as input which are usually unavailable. To quantitatively evaluate how the advisor-advisee relationships can help bole search, we compare a retrieval method with and without those relationships on a data set used in Yang et al. [169]. Specifically, the data set consists 9 queries (e.g., data mining and machine learning), and for each query, 50 top-ranked researchers by ArnetMiner.org are taken as candidates. We sent an email to each of the 50 researchers and another 50 young researchers who start publishing papers only in recent years (> 2003) for feedbacks ("yes," or "no," or "not sure"). Finally a list of best supervisors are organized for each query by simply counting the number of "yes"(+1) and "no"(-1) from the 100 received feedbacks. Details can be referred to in Yang et al. [169]. For easy comparison, we did not use the learning-to-rank approach (as reported in Yang et al. [169]). Instead, we use the language model (LM), which does not consider the advisor-advisee relationships, and a heuristic-based method which simply combines the language model with the advisor-advisee relationships identified by the baseline method (LM+Rule) or identified our proposed approach (LM+TPFG) by

$$s_i = \alpha r_i + (1 - \alpha) \frac{1}{|A_i|} \sum_{a_j \in A_i} r_{a_j}, \tag{5.28}$$

where r_i is the relevance score obtained by the language model; A_i is a set of advisees of researcher a_i identified by the rule-based method or TPFG; and α is a parameter to trade off the balance between researcher's expertise and his advisees' expertise score. We empirically set $\alpha = 0.7$.

Figure 5.13(b) shows the results (Precision@2, Precision@5, mean average precision (MAP), and NDCG@5 [7]) of bole search by the three methods. We see that by considering

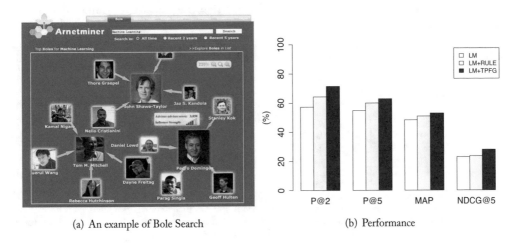

(a) An example of Bole Search   (b) Performance

Figure 5.13:   Application: from Expert Finding to Bole Search.

the advisor-advisee relationships (obtained by either rule-based method or TPFG) the performance of bole search can be significantly improved. We can also see that with a higher accuracy,
TPFG clearly achieves a better improvements, particularly for the top two retrieved results (71.4%
by TPFG vs. 64.3% by rule-based method in terms of P@2).

5.6.2   MEASURING ACADEMIC INFLUENCE

We applied the Topical Affinity Propagation (TAP) method presented in Section 3 to Arnet-
Miner to identify representative authors and representative papers.

Table 5.13 shows representative nodes (authors and papers) found by our algorithm on
different topics from the coauthor data set and the citation data set. The representative score of
each node is the probability of the node influencing the other nodes on this topic. The probability
is calculated by $\frac{\sum_{j \in NB(i) \cup \{i\}} \mu_{ij}}{\sum_{i=1}^{N} \sum_{j \in NB(i) \cup \{j\}} \mu_{ij}}$. We can see some interesting results. For example, some
papers (e.g., "FaCT and iFaCT") that do have have a high citation number might be selected as
the representative nodes. This is because our algorithm can identify the influences between papers,
thus can differentiate the citations of the theoretical background of a paper and an odd citation
in the reference.

Table 5.14 shows four representative authors and researchers who are mostly influenced by
them. Table 5.15 shows two representative papers and papers that are mostly influence by the
two papers. Some other method, e.g., the similarity-based baseline method using cosine metric,
can be also used to estimate the influence according to the similarity score. Comparing with the
similarity-based baseline method, the presented TAP method has several distinct advantages.
First, such a method can only measure the similarity between nodes, but cannot tell which node
has a stronger influence on the other one. Second, the method cannot tell which nodes have the

Table 5.13: Representative nodes discovered by our algorithm on the co-author data set and the citation data set

Dataset	Topic	Representative Nodes
Author	Data Mining	Heikki Mannila, Philip S. Yu, Dimitrios Gunopulos, Jiawei Han, Christos Faloutsos, Bing Liu, Vipin Kumar, Tom M. Mitchell, Wei Wang, Qiang Yang, Xindong Wu, Jeffrey Xu Yu, Osmar R. Zaiane
	Machine Learning	Pat Langley, Alex Waibel, Trevor Darrell, C. Lee Giles, Terrence J. Sejnowski, Samy Bengio, Daphne Koller, Luc De Raedt, Vasant Honavar, Floriana Esposito, Bernhard Scholkopf
	Database System	Gerhard Weikum, John Mylopoulos, Michael Stonebraker, Barbara Pernici, Philip S. Yu, Sharad Mehrotra, Wei Sun, V. S. Subrahmanian, Alejandro P. Buchmann, Kian-Lee Tan, Jiawei Han
	Information Retrieval	Gerard Salton, W. Bruce Croft, Ricardo A. Baeza-Yates, James Allan, Yi Zhang, Mounia Lalmas, Zheng Chen, Ophir Frieder, Alan F. Smeaton, Rong Jin
	Web Services	Yan Wang, Liang-jie Zhang, Schahram Dustdar, Jian Yang, Fabio Casati, Wei Xu, Zakaria Maamar, Ying Li, Xin Zhang, Boualem Benatallah, Boualem Benatallah
	Semantic Web	Wolfgang Nejdl, Daniel Schwabe, Steffen Staab, Mark A. Musen, Andrew Tomkins, Juliana Freire, Carole A. Goble, James A. Hendler, Rudi Studer, Enrico Motta
	Bayesian Network	Daphne Koller, Paul R. Cohen, Floriana Esposito, Henri Prade, Michael I. Jordan, Didier Dubois, David Heckerman, Philippe Smets
Citation	Data Mining	Fast Algorithms for Mining Association Rules in Large Databases, Using Segmented Right-Deep Trees for the Execution of Pipelined Hash Joins, Web Usage Mining: Discovery and Applications of Usage Patterns from Web Data, Discovery of Multiple-Level Association Rules from Large Databases, Interleaving a Join Sequence with Semijoins in Distributed Query Processing
	Machine Learning	Object Recognition with Gradient-Based Learning, Correctness of Local Probability Propagation in Graphical Models with Loops, A Learning Theorem for Networks at Detailed Stochastic Equilibrium, The Power of Amnesia: Learning Probabilistic Automata with Variable Memory Length, A Unifying Review of Linear Gaussian Models
	Database System	Mediators in the Architecture of Future Information Systems, Database Techniques for the World-Wide Web: A Survey, The R*-Tree: An Efficient and Robust Access Method for Points and Rectangles, Fast Algorithms for Mining Association Rules in Large Databases
	Web Services	The Web Service Modeling Framework WSMF, Interval Timed Coloured Petri Nets and their Analysis, The design and implementation of real-time schedulers in RED-linux, The Self-Serv Environment for Web Services Composition
	Web Mining	Web Usage Mining: Discovery and Applications of Usage Patterns from Web Data, Fast Algorithms for Mining Association Rules in Large Databases, The OO-Binary Relationship Model: A Truly Object Oriented Conceptual Model, Distributions of Surfers' Paths Through the World Wide Web: Empirical Characterizations, Improving Fault Tolerance and Supporting Partial Writes in Structured Coterie Protocols for Replicated Objects
	Semantic Web	FaCT and iFaCT, The GRAIL concept modelling language for medical terminology, Semantic Integration of Semistructured and Structured Data Sources, Description of the RACER System and its Applications, DL-Lite: Practical Reasoning for Rich Dls

highest influences in the network, which the presented TAP approach naturally has the capacity to do. This provides many immediate applications, for example, expert finding.

We further conduct a dynamic influence analysis. We use Dr. Jian Pei as the example to analyze how the influences of Dr. Pei on or by his coauthors change during 2000 and 2008. Table 5.16 shows the dynamic analysis result. We see that the influence evolution uncovers the growing up of Dr. Pei. For example, in 2000 Dr. Pei is mainly influenced by Prof. Han, while he only has limited influence on Prof. Han. After 2004, Dr. Pei starts influencing some other researchers (e.g., Chun Tang and Shiwei Tang). While in 2008, Dr. Pei already becomes a mature researcher and has many strong influences on other researchers.

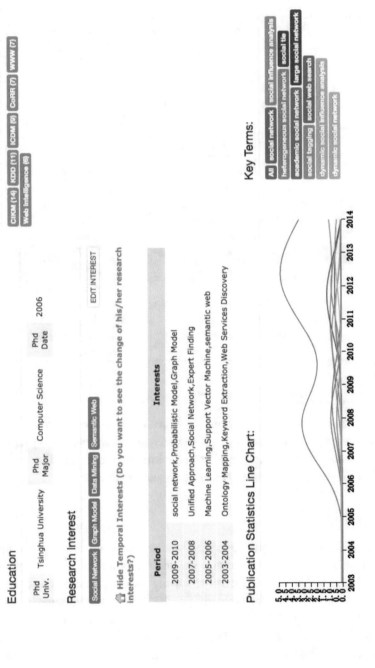

Figure 5.14: An example result of user interest modeling.

Table 5.14: Example of influence analysis from the coauthor data set. There are two representative authors and example list of researchers who are mostly influenced by them on topic "data mining," and their corresponding influenced order on topic "database" and "machine learning"

Topic	Data Mining	
Influencer	Jiawei Han	Heikki Mannila
Influencee	David Clutter Hasan M. Jamil K. P. Unnikrishnan Ramasamy Uthurusamy Shiwei Tang	Arianna Gallo Marcel Holsheimer Robert Gwadera Vladimir Estivill-Castro Mika Klemettinen
Topic	Database	
Influencer	Jiawei Han	Heikki Mannila
Influencee	David Clutter Shiwei Tang Hasan M. Jamil Ramasamy Uthurusamy K. P. Unnikrishnan	Vladimir Estivill-Castro Marcel Holsheimer Robert Gwadera Mika Klemettinen Arianna Gallo
Topic	Machine Learning	
Influencer	Jiawei Han	Heikki Mannila
Influencee	Hasan M. Jamil K. P. Unnikrishnan Shiwei Tang Ramasamy Uthurusamy David Clutter	Vladimir Estivill-Castro Marcel Holsheimer Mika Klemettinen Robert Gwadera Arianna Gallo

5.6.3 MODELING RESEARCHER INTERESTS

We applied the method presented in Section 4 to ArnetMiner to model authors' research interests. The basic idea is to use the prediction model to predict who will be interested in which (predefined) topics. Figure 5.14 shows an example result of user interest modeling. The middle of the figure shows the author's research interests and interest changes over time.

5.7 CONCLUSIONS

In this chapter, we introduce a system—ArnetMiner—for deep mining of a researcher social network. We introduce the architecture and the main features of the system. We describe in detail the research issues that we are currently focusing on and proposed our approaches to them. We have carried out experiments for evaluating each of the proposed approaches. Experimental results indicate that the proposed methods can achieve high performances.

In the implementation of ArnetMiner, we have a few problems which need to be investigated in the future, for example, extraction of more types of relationships. In ArnetMiner, we use

Table 5.15: Example of influence analysis results on topic "data mining" from the citation data set. There are two representative papers and example paper lists that are mostly influenced by them

Influential paper	Fast Algorithms for Mining Association Rules in Large Databases
Influenced paper	Mining Large Itemsets for Association Rules A New Framework For Itemset Generation Efficient Mining of Partial Periodic Patterns in Time Series Database A New Method for Similarity Indexing of Market Basket Data A General Incremental Technique for Maintaining Discovered Association Rules
Influential paper	Web Usage Mining: Discovery and Applications of Usage Patterns from Web Data
Influenced paper	Mining Web Site's Clusters from Link Topology and Site Hierarchy Predictive Algorithms for Browser Support of Habitual User Activities on the Web A Fine Grained Heuristic to Capture Web Navigation Patterns A Road Map to More Effective Web Personalization: Integrating Domain Knowledge with Web Usage Mining

Table 5.16: Dynamic influence analysis for Dr. Jian Pei during 2000–2008. Due to space limitation, we only list coauthors who most influence on/by Dr. Pei in each time window

Year	Pairwise	Influence
2000	Influence on Dr. Pei	Jiawei Han (0.4961)
	Influenced by Dr. Pei	Jiawei Han (0.0082)
2002	Influence on Dr. Pei	Jiawei Han (0.4045), Ke Wang (0.0418), Jianyong Wang (0.019), Xifeng Yan (0.007), Shiwei Tang (0.0052)
	Influenced by Dr. Pei	Shiwei Tang (0.436), Hasan M.Jamil (0.4289), Xifeng Yan (0.2192), Jianyong Wang (0.1667), Ke Wang (0.0687)
2004	Influence on Dr. Pei	Jiawei Han (0.2364), Ke Wang (0.0328), Wei Wang (0.0294), Jianyong Wang (0.0248), Philip S. Yu (0.0156)
	Influenced by Dr. Pei	Chun Tang (0.5929), Shiwei Tang (0.5426), Hasan M.Jamil (0.3318), Jianyong Wang (0.1609), Xifeng Yan (0.1458)
2006	Influence on Dr. Pei	Jiawei Han (0.1201), Ke Wang (0.0351), Wei Wang (0.0226), Jianyong Wang (0.018), Ada Wai-Chee Fu (0.0125)
	Influenced by Dr. Pei	Chun Tang (0.6095), Shiwei Tang (0.6067), Byung-Won On (0.4599), Hasan M.Jamil (0.3433), Jaewoo Kang (0.3386)
2008	Influence on Dr. Pei	Jiawei Han (0.2202), Ke Wang (0.0234), Ada Wai-Chee Fu (0.0208), Wei Wang (0.011), Jianyong Wang (0.0095)
	Influenced by Dr. Pei	ZhaoHui Tang (0.654), Chun Tang (0.6494), Shiwei Tang (0.5923), Zhengzheng Xing (0.5549), Hasan M.Jamil (0.3333)

the co-author relationship as the relationship in a social network. In the future, we will extract other relationships, for example, the relationship of co-organization and co-project, etc. In future work, we also plan to investigate more mining issues, for example how to integrate the citation information into the topic models.

CHAPTER 6

Research Frontiers

The general problem of mining knowledge links and social networks represents an interesting research direction in computer science. There are still many challenges and also potential future directions on this topic. Here we list several major challenges.

- *Big data vs. small data.* Big data is one of the hottest terms used today to describe the exponential growth and availability of data on the Internet. Technically, there are several challenges and also opportunities. First, as the data is generated dynamically and arrives in a streaming manner, it is necessary to develop efficient algorithm for incrementally mining the semantics of links and social relationships in large dynamic networks. On the other hand, people also argue that big data may be not that important. In many practical applications, small data might be sufficient to achieve an enough accurate performance. Big data may only introduce extra computational cost. The challenging question is: when should we resort to big data and when should we simply consider sampled small data.

- *Macro vs. micro.* At the macro-level, social network mining focuses on studying the global patterns of the social networks; and at the micro-level, it is mainly concerned with modeling individuals' behaviors and interactions between users. What is the fundamental relationship between the two types of analyses? When should we consider macro-level analysis and when micro?

- *Globality vs. locality.* Most existing works focus on studying social relationships in the entire network. However, as most users and their behaviors are influenced by friends in their local circles. It would be interesting to study how semantics of social relationships can be inferred from the locality perspective. There are some related research on this topic. For example, McAuley and Leskovec developed a semi-supervised learning framework for automatically grouping a user's friends into different circles [104]. Zhang et al. proposed the notion of social influence locality [174]. They find that the influence between users is strongly correlated with the inner structure formed by them.

- *Correlation vs. causality.* Most existing methodologies only consider correlation between variables. However, correlation is not equal to causality. In many problems, we may be more interested in understanding the reason of a phenomenon rather than observing other correlated phenomena.

- *Precise vs. approximation.* To precisely process the networked data is computationally costly. However, as online networks become larger and larger, it is necessary to find some efficient algorithms to speed up the process even with some loss on the precision. From the algorithmic perspective, it is challenging to design effective algorithms with theoretical approximation guarantees.

- *Social theories.* How does one seamlessly incorporate social theories into the mining algorithms? Traditionally, this problem has been dealt with in an ad hoc manner. For example, Hopcroft et al. [70] and Tang et al. [142] combined social balance theory, social status theory, and structural hole theory into the factor graph model. However, it is still not clear how to develop a unified model so that the other social theories can be easily incorporated. More importantly, the social networks are very dynamic. The semantic of social relationships may change over time. It is important to capture the dynamic pattern and infer the changes of social relationships.

- *Applications.* It has many real applications based on the results of mining knowledge links and social tie. For example, we can use the inferred social ties to help information recommendation in the social network. According to the social influence theory, a user's connections with different social ties would have very different influence on her/his behaviors from different aspects.

APPENDIX A

Resources

We have developed a Social Analytic Engine (SAE) for analyzing and mining large social network. The cornerstone of the analytic engine is a distributed graph database, which provides storage for the networking data. On top of the database, there are three core components: network integration, social network analysis, and distributed machine learning. The toolkit can be download here:

`https://github.com/actnet/saedb.`

A.1 SOFTWARE

We also provide a list of softwares various social tie analysis, social influence analysis and user behavior prediction.

- **PLP-FGM**: The PLP-FGM model [150] is designed for inferring the type of social relationships in a partially labeled social network.
  `http://arnetminer.org/socialtie`

- **TriFG**: The Triad Factor Graph model [70, 101] aims to predict the formation of reciprocal relationships and the formation of triadic closure.
  `http://arnetminer.org/reciprocity`

- **TAP**: The Topical Affinity Propagation algorithm [143] is to estimate the topic-level social influence on large networks.
  `http://arnetminer.org/lab-datasets/soinf/`

- **Confluence**: The Confluence model [144] formalizes the effects of social conformity into a probabilistic factor graph model and can distinguish and quantify the effect of the different types of conformities.
  `http://arnetminer.org/Confluence`

- **NTT-FGM**: The NTT-FGM model [138] formalizes social influence, correlation (homophily), and users' action dependency into a unified approach and distinguishes their effects for modeling and predicting users' actions in social networks.
  `http://arnetminer.org/stnt/`

- **ACT**: The Author-Conference-Topic [148] model can simultaneously model topical distributions of papers, authors, and conferences (or any three different types of entities).
  `http://keg.cs.tsinghua.edu.cn/jietang/software/ACT-code.zip`

- **HIS&MaxD**: The two algorithms aim to detect top-k structural hole spanners in large-scale social networks. With the algorithms, we discovered that 1who span structural holes control 25% of the information diffusion on Twitter.
  `http://arnetminer.org/structural-hole`

- **RAIN**: The model can jointly model users roles and information diffusion. It can discover different social roles and also model how users of different social roles control the information diffusion.
  `http://aminer.org/billboard/rain`

- **WhoAmI**: The WhoAmI model aims to predict user demographics based on their communication behaviors.
  `http://arnetminer.org/demographic`

A.2   DATA SETS

We also provide a list of data sets for research on social networks.

- **Citation**: The citation data is extracted from DBLP, ACM, and other sources, consisting of >2,000,000 papers and >4,000,000 citation relationships. The data set can be used for clustering with network and side information, studying influence in the citation network, finding the most influential papers, topic modeling analysis, etc.
  `http://arnetminer.org/citation`
  `http://arnetminer.org/AMinerNetwork` (containing more information related to authors).

- **Weibo**: This is a dynamic Weibo networking data, containing more than 1,700,000 users, 308,000,000 following relationships, 300,000 original tweets and 23,000,000 retweets. The dataset is unique, as it contains all the dynamic information.
  `http://arnetminer.org/Influencelocality`

- **Tencent Weibo**: The dataset contains the directed following networks and posting logs of over 200 million users in Tencent Weibo. In total, we have 184,491 users, and 4,588,559 original posts. We removed original posts that were reposted by fewer than 5 users, and use the remaining 242,831 original posts for experiments.
  `http://arnetminer.org/rain`

- **Flickr**: The Flickr dataset consists of 354,192 randomly downloaded images posted by 4,807 users. For each image, we also collect its tags and all comments. In total, we have 557,177 comments posted by 6,735 users.

`http://arnetminer.org/emotion`

- **Name disambiguation**: The data set is from Arnetminer.org, consisting of manually labeled disambiguation results for 6,730 papers authored by 100 author names.

  `http://arnetminer.org/disambiguation`

Bibliography

[1] Lada A. Adamic and Eytan Adar. Friends and neighbors on the web. *Social Networks*, 25:211–230, 2001. DOI: 10.1016/S0378-8733(03)00009-1. 16

[2] Aris Anagnostopoulos, Ravi Kumar, and Mohammad Mahdian. Influence and correlation in social networks. In *KDD'08*, pages 7–15, 2008. DOI: 10.1145/1401890.1401897. 6, 69, 90, 111

[3] Christophe Andrieu, Nando de Freitas, Arnaud Doucet, and Michael I. Jordan. An introduction to mcmc for machine learning. *Machine Learning*, 50:5–43, 2003. DOI: 10.1023/A:1020281327116. 150

[4] Lars Backstrom, Dan Huttenlocher, Jon Kleinberg, and Xiangyang Lan. Group formation in large social networks: membership, growth, and evolution. In *KDD'06*, pages 44–54, 2006. DOI: 10.1145/1150402.1150412. 2

[5] Lars Backstrom, Ravi Kumar, Cameron Marlow, Jasmine Novak, and Andrew Tomkins. Preferential behavior in online groups. In *WSDM'08*, pages 117–128, 2008. DOI: 10.1145/1341531.1341549. 8, 111

[6] Lars Backstrom and Jure Leskovec. Supervised random walks: predicting and recommending links in social networks. In *WSDM'11*, pages 635–644, 2011. DOI: 10.1145/1935826.1935914. 5, 13, 15, 40

[7] Ricardo Baeza-Yates and Berthier Ribeiro-Neto. *Modern Information Retrieval*. ACM Press, 1999. 155, 164

[8] Norman T.J Bailey. *The Mathematical Theory of Infectious Diseases and its Applications*. LondonHigh Wycombe: Charles Griffin, 1975. 118

[9] Eytan Bakshy, Dean Eckles, Rong Yan, and Itamar Rosenn. Social influence in social advertising: evidence from field experiments. In *EC'12*, pages 146–161, 2012. DOI: 10.1145/2229012.2229027. 2, 69

[10] Albert-László Barabási and Réka Albert. Emergence of scaling in random networks. *Science*, 286(5439):509–512, 1999. DOI: 10.1126/science.286.5439.509. 2

[11] Sugato Basu, Mikhail Bilenko, and Raymond J. Mooney. A probabilistic framework for semi-supervised clustering. In *Proceedings of the Tenth ACM SIGKDD International Conference on Knowledge Discovery and Data Mining (KDD'04)*, pages 59–68, 2004. DOI: 10.1145/1014052.1014062. 141

[12] Václav Belak, Samantha Lam, and Conor Hayes. Cross-community influence in discussion fora. In *ICWSM'12*, pages 34–41, 2012. 7

[13] David M. Blei and Jon D. McAuliffe. Supervised topic models. In *Proceedings of the 19th Neural Information Processing Systems (NIPS'07)*, 2007. 153

[14] David M. Blei, Andrew Y. Ng, and Michael I. Jordan. Latent dirichlet allocation. *JMLR*, 3:993–1022, 2003. 71, 148, 150, 155

[15] Philip Bonacich. Factoring and weighting approaches to status scores and clique identification. *Journal of Mathematical Sociology*, 2(1):113–120, 1972. DOI: 10.1080/0022250X.1972.9989806. 99

[16] Philip Bonacich. Power and centrality: A family of measures. *The American Journal of Sociology*, 92(5):1170–1182, 1987. DOI: 10.1086/228631. 98

[17] Philip Bonacich and Paulette Lloyd. Eigenvector-like measures of centrality for asymmetric relations. *Social Networks*, 23(3):191–201, 2001. DOI: 10.1016/S0378-8733(01)00038-7. 98

[18] Robert M. Bond, Christopher J. Fariss, Jason J. Jones, Adam D. I. Kramer, Cameron Marlow, Jaime E. Settle, and James H. Fowler. A 61-million-person experiment in social influence and political mobilization. *Nature*, 489:295–298, 2012. DOI: 10.1038/nature11421. 6, 69

[19] Dan Brickley and Libby Miller. Foaf vocabulary specification. In *Namespace Document, http://xmlns.com/foaf/0.1/*, September 2004. 132, 134

[20] Chris Buckley and Ellen M. Voorhees. Retrieval evaluation with incomplete information. In *SIGIR'2004*, pages 25–32, 2004. DOI: 10.1145/1008992.1009000. 85, 155

[21] Ronald S. Burt. *Structural Holes: The Social Structure of Competition*. Harvard University Press, 1992. 2, 3, 53, 54

[22] Deng Cai, Xiaofei He, and Jiawei Han. Spectral regression for dimensionality reduction. In *Technical Report. UIUCDCS-R-2007-2856, UIUC*, 2007. 146

[23] Mary Elaine Califf and Raymond J. Mooney. Relational learning of pattern-match rules for information extraction. In *Proceedings of Association for the Advancement of Artificial Intelligence (AAAI'99)*, pages 328–334, 1999. 6, 40

[24] Jeremy J. Carroll, Ian Dickinson, Chris Dollin, Dave Reynolds, Andy Seaborne, and Kevin Wilkinson. Jena: Implementing the semantic web recommendations. In *WWW'04*, pages 74–83, 2004. DOI: 10.1145/1013367.1013381. 132

[25] Meeyoung Cha, Alan Mislove, and Krishna P. Gummadi. A measurement-driven analysis of information propagation in the flickr social network. In *WWW'09*, pages 367–376, 2009. DOI: 10.1145/1526709.1526806. 123

[26] Wei Chen, Chi Wang, and Yajun Wang. Scalable influence maximization for prevalent viral marketing in large-scale social networks. In *KDD'10*, pages 1029–1038, 2010. DOI: 10.1145/1835804.1835934. 7, 82, 86

[27] Wei Chen, Yajun Wang, and Siyu Yang. Efficient influence maximization in social networks. In *KDD'09*, pages 199–207, 2009. DOI: 10.1145/1557019.1557047. 7, 70, 82, 86

[28] Fabio Ciravegna. An adaptive algorithm for information extraction from web-related texts. In *Proceedings of the IJCAI-2001 Workshop on Adaptive Text Extraction and Mining*, August 2001. 138

[29] Aaron Clauset, Cristopher Moore, and M. E. J. Newman. Hierarchical structure and the prediction of missing links in networks. *Nature*, 453(7191):98–101, May 2008. DOI: 10.1038/nature06830. 12, 13

[30] Corinna Cortes and Vladimir Vapnik. Support-vector networks. *Machine Learning*, 20:273–297, 1995. DOI: 10.1023/A:1022627411411. 136, 138

[31] David Crandall, Dan Cosley, Daniel Huttenlocher, Jon Kleinberg, and Siddharth Suri. Feedback effects between similarity and social influence in online communities. In *KDD'08*, pages 160–168, 2008. DOI: 10.1145/1401890.1401914. 7, 69, 90, 111

[32] David J. Crandall, Lars Backstrom, Dan Cosley, Daniel Huttenlocher Siddharth Suri and, and Jon Kleinberg. Inferring social ties from geographic coincidences. *PNAS*, 107:22436–22441, December 2010. 40

[33] Nick Craswell, Arjen P. de Vries, and Ian Soboroff. Overview of the trec-2005 enterprise track. In *TREC 2005 Conference Notebook*, pages 199–205, 2005. 85, 155

[34] James A. Davis and Samuel Leinhardt. The structure of positive interpersonal relations in small groups. In J. Berger, editor, *Sociological Theories in Progress*, vol. 2, pages 218–251. Houghton Mifflin, 1972. 2, 3, 53, 55

[35] Abir De, Niloy Ganguly, and Soumen Chakrabarti. Discriminative link prediction using local links, node features and community structure. In *ICDM*, pages 1009–1018, 2013. DOI: 10.1109/ICDM.2013.68. 13

[36] Jeffrey Dean and Sanjay Ghemawat. Mapreduce: Simplified data processing on large clusters. In *OSDI'04*, pages 10–10, 2004. DOI: 10.1145/1327452.1327492. 77

[37] M. Deutsch and H. B. Gerard. A study of normative and informational social influences upon individual judgment. *Journal of Abnormal and Social Psychology*, 51:629–636, 1955. DOI: 10.1037/h0046408. 69

[38] Christopher P. Diehl, Galileo Namata, and Lise Getoor. Relationship identification for social network discovery. In *AAAI*, pages 546–552, 2007. 5, 25, 35, 40, 52

[39] Laura Dietz, Steffen Bickel, and Tobias Scheffer. Unsupervised prediction of citation influences. In *ICML'07*, pages 233–240, 2007. DOI: 10.1145/1273496.1273526. 101, 102

[40] Edsger W. Dijkstra. A note on two problems in connexion with graphs. *Numerische Mathematik*, 1:269–271, 1959. DOI: 10.1007/BF01386390. 163

[41] Peter Sheridan Dodds, Roby Muhamad, and Duncan J Watts. An experimental study of search in global social networks. *Science*, 301(5634):827–829, 2003. DOI: 10.1126/science.1081058. 2

[42] Pedro Domingos and Matt Richardson. Mining the network value of customers. In *KDD'01*, pages 57–66, 2001. DOI: 10.1145/502512.502525. 7, 70, 85, 111

[43] Yon Dourisboure, Filippo Geraci, and Marco Pellegrini. Extraction and classification of dense communities in the web. In *WWW'2007*, pages 461–470, 2007. DOI: 10.1145/1242572.1242635. 7

[44] Nathan Eagle, Alex (Sandy) Pentland, and David Lazer. Inferring social network structure using mobile phone data. *PNAS*, 106(36), 2009. DOI: 10.1007/978-0-387-77672-9_10. 5, 35, 52

[45] David Easley and Jon Kleinberg. *Networks, Crowds, and Markets: Reasoning about a Highly Connected World*. Cambridge University Press, 2010. DOI: 10.1017/CBO9780511761942. 3, 53

[46] P. Erdos and A. Renyi. On the evolution of random graphs. *Publications of the Mathematical Institute of the Hungarian Academy of Sciences*, 5:17–61, 1960. 2

[47] Michalis Faloutsos, Petros Faloutsos, and Christos Faloutsos. On power-law relationships of the internet topology. In *SIGCOMM'99*, pages 251–262, 1999. DOI: 10.1145/316194.316229. 2

[48] Gerhard Fischer. User odeling in human–computer interaction. *User Modeling and User-Adapted Interaction*, 11(1-2):65–86, 2001. DOI: 10.1023/A:1011145532042. 2, 8, 111

[49] James H. Fowler and Nicholas A. Christakis. Dynamic spread of happiness in a large social network: longitudinal analysis over 20 years in the framingham heart study. In *British Medical Journal*, 2008. DOI: 10.1136/bmj.a2338. 69

[50] Brendan J. Frey and Delbert Dueck. Mixture modeling by affinity propagation. In *Proceedings of the 18th Neural Information Processing Systems (NIPS'06)*, pages 379–386, 2006. 8

[51] Lise Getoor and Ben Taskar. *Introduction to Statistical Relational Learning*. The MIT Press, 2007. 6

[52] Zoubin Ghahramani and Michael I. Jordan. Factorial hidden markov models. *Machine Learning*, 29(2-3):245–273, 1997. DOI: 10.1023/A:1007425814087. 115, 116

[53] M. Girvan and M. E. J. Newman. Community structure in social and biological networks. *PNAS*, 99(12):7821–7826, 2002. 2

[54] Amit Goyal, Francesco Bonchi, and Laks V. S. Lakshmanan. A data-based approach to social influence maximization. *Proc. VLDB Endow.*, 5(1):73–84, 2012. DOI: 10.14778/2047485.2047492. 70

[55] Amit Goyal, Francesco Bonchi, and Laks V.S. Lakshmanan. Learning influence probabilities in social networks. In *WSDM'10*, pages 241–250, 2010. DOI: 10.1145/1718487.1718518. 7, 70

[56] Mark Granovetter. The strength of weak ties. *American Journal of Sociology*, 78(6):1360–1380, 1973. DOI: 10.1086/225469. 2, 3, 4, 89

[57] Thomas L. Griffiths and Mark Steyvers. Finding scientific topics. In *PNAS'04*, pages 5228–5235, 2004. DOI: 10.1073/pnas.0307752101. 150

[58] Reto Grob, Michael Kuhn, Roger Wattenhofer, and Martin Wirz. Cluestr: Mobile social networking for enhanced group communication. In *GROUP'09*, 2009. DOI: 10.1145/1531674.1531686. 11

[59] Daniel Gruhl, R. Guha, David Liben-Nowell, and Andrew Tomkins. Information diffusion through blogspace. In *WWW'04*, pages 491–501, 2004. DOI: 10.1145/1046456.1046462. 2, 7, 111, 118

[60] R. Guha, Ravi Kumar, Prabhakar Raghavan, and Andrew Tomkins. Propagation of trust and distrust. In *WWW'04*, pages 403–412, 2004. DOI: 10.1145/988672.988727. 3, 53, 55, 96

[61] Roger Guimerà and Marta Sales-Pardo. Missing and spurious interactions and the reconstruction of complex networks. *Proceedings of the National Academy of Sciences*, 106:22073–22078, 2009. DOI: 10.1073/pnas.0908366106. 13, 14

[62] Honglei Guo, Huijia Zhu, Zhili Guo, XiaoXun Zhang, and Zhong Su. Address standardization with latent semantic association. In *KDD'09*, pages 1155–1164, 2009. DOI: 10.1145/1557019.1557144. 111

[63] J. M. Hammersley and P. Clifford. Markov field on finite graphs and lattices. *Unpublished manuscript*, 1971. 31, 58

[64] Mohammad Al Hasan and Mohammed J. Zaki. A survey of link prediction in social networks. In *Social Network Data Analytics*, pages 243–275. Kluwer Academic Publishers, 2011. DOI: 10.1007/978-1-4419-8462-3_9. 15, 18, 20

[65] Taher H. Haveliwala. Topic-sensitive pagerank. In *Proceedings of the 11th International Conference on World Wide Web (WWW'02)*, pages 517–526, 2002. DOI: 10.1109/TKDE.2003.1208999. 108

[66] Simon S. Haykin. *Kalman Filtering and Neural Networks*. John Wiley & Sons, Inc., New York, 2001. DOI: 10.1002/0471221546. 115, 116

[67] Creighton Heaukulani and Zoubin Ghahramani. Dynamic probabilistic models for latent feature propagation in social networks. In *ICML'13*, volume 28, pages 275–283, 2013. 13, 15

[68] Georey E. Hinton. A fast learning algorithm for deep belief nets. *Neural Computation*, 18:1527–1554, 2006. DOI: 10.1162/neco.2006.18.7.1527. 8

[69] Thomas Hofmann. Probabilistic latent semantic indexing. In *SIGIR'99*, pages 50–57, 1999. DOI: 10.1145/312624.312649. 71, 148

[70] John Hopcroft, Tiancheng Lou, and Jie Tang. Who will follow you back? reciprocal relationship prediction. In *CIKM'11*, pages 1137–1146, 2011. DOI: 10.1145/2063576.2063740. 5, 11, 41, 172

[71] Glen Jeh and Jennifer Widom. Scaling personalized web search. In *WWW'02*, pages 271–279, 2002. DOI: 10.1145/775152.775191. 97

[72] F. V. Jensen. *An Introduction to Bayesian Networks*. Springer-Verlag, New York, 1996. 8

[73] George Karypis and Vipin Kumar. *MeTis: Unstrctured Graph Partitioning and Sparse Matrix Ordering System, Version 4.0*, September 1998. 33

[74] Hisashi Kashima and Naoki Abe. A parameterized probabilistic model of network evolution for supervised link prediction. In *ICDM*, pages 340–349, 2006. DOI: 10.1109/ICDM.2006.8. 12, 13

[75] Elihu Katz. The two-step flow of communication: an up-to-date report of an hypothesis. In Enis and Cox (eds.), *Marketing Classics*, pages 175–193, 1973. 4

[76] Elihu Katz and Paul Felix Lazarsfeld. *Personal Influence*. The Free Press, New York, 1955. 4

[77] Leo Katz. A new status index derived from sociometric analysis. *Psychometrika*, 18(1):39–43, 1953. DOI: 10.1007/BF02289026. 99

[78] H. C Kelman.   Compliance, identification, and internalization: Three processes of attitude change.   *Journal of Conflict Resolution*, 2(1):51–60, 1958. DOI: 10.1177/002200275800200106. 6, 69

[79] David Kempe, Jon Kleinberg, and Éva Tardos. Maximizing the spread of influence through a social network. In *KDD'03*, pages 137–146, 2003. DOI: 10.1145/956750.956769. 2, 7, 11, 42, 43, 44, 70, 82, 85, 111

[80] Masahiro Kimura and Kazumi Saito. Tractable models for information diffusion in social networks. In *PKDD'06*, pages 259–271, 2006. DOI: 10.1007/11871637_27. 86

[81] Christine Kiss and Martin Bichler.   Identification of influencers- measuring influence in customer networks.   *Decision Support Systems*, 46(1):233–253, 2008. DOI: 10.1016/j.dss.2008.06.007. 99

[82] Jon Kleinberg. Temporal dynamics of on-line information streams. In *Data Stream Managemnt: Processing High-speed Data*. Springer, 2005. 111

[83] David Krackhardt. *The Strength of Strong Ties: The Importance of Philos in Networks and Organization*, In Nitin Nohria and Robert G. Eccles, (Eds.), Networks and Organizations. Harvard Business School Press, Cambridge, 1992. 3, 89

[84] Frank R. Kschischang, Brendan J. Frey, and Hans andrea Loeliger. Factor graphs and the sum-product algorithm. *IEEE TOIT*, 47:498–519, 2001. DOI: 10.1109/18.910572. 8, 72, 75

[85] Jérôme Kunegis and Jörg Fliege. Predicting directed links using nondiagonal matrix decomposition. In *ICDM'12*, pages 948–953, 2012. DOI: 10.1109/ICDM.2012.16. 13, 14

[86] Jérôme Kunegis and Andreas Lommatzsch. Learning spectral graph transformations for link prediction. In *ICML*, pages 561–568, 2009. DOI: 10.1145/1553374.1553447. 13, 14, 20, 23, 24

[87] John D. Lafferty, Andrew McCallum, and Fernando C. N. Pereira. Conditional random fields: Probabilistic models for segmenting and labeling sequence data. In *ICML'01*, pages 282–289, 2001. 31, 61, 116, 136, 137

[88] P. F. Lazarsfeld and R. K. Merton. Friendship as a social process: A substantive and methodological analysis. In M. Berger, T. Abel, and C. H. Page, (Eds.), *Freedom and Control in Modern Society*, Van Nostrand, New York, pages 8–66, 1954. 11

[89] Paul Felix Lazarsfeld, Bernard Berelson, and Hazel Gaudet. *The People's Choice: How the Voter Makes up His Mind in a Presidential Campaign*. Columbia University Press, New York, 1944. 3, 4, 53, 56

[90] Conrad Lee, Bobo Nick, Ulrik Brandes, and Pádraig Cunningham. Link prediction with social vector clocks. In *KDD*, pages 784–792, 2013. DOI: 10.1145/2487575.2487615. 13

[91] Vincent Leroy, Berkant Barla Cambazoglu, and Francesco Bonchi. Cold start link prediction. In *KDD*, pages 393–402, 2010. DOI: 10.1145/1835804.1835855. 13, 14

[92] Jure Leskovec, Daniel Huttenlocher, and Jon Kleinberg. Predicting positive and negative links in online social networks. In *WWW'10*, pages 641–650, 2010. DOI: 10.1145/1772690.1772756. 5, 41, 61

[93] Jure Leskovec, Daniel Huttenlocher, and Jon Kleinberg. Signed networks in social media. In *CHI'10*, pages 1361–1370, 2010. DOI: 10.1145/1753326.1753532. 3, 53, 55

[94] Jure Leskovec, Andreas Krause, Carlos Guestrin, Christos Faloutsos, Jeanne VanBriesen, and Natalie Glance. Cost-effective outbreak detection in networks. In *KDD'07*, pages 420–429, 2007. DOI: 10.1145/1281192.1281239. 7

[95] Jure Leskovec, Kevin J. Lang, Anirban Dasgupta, and Michael W. Mahoney. Statistical properties of community structure in large social and information networks. In *WWW'08*, pages 695–704, 2008. DOI: 10.1145/1367497.1367591. 111

[96] David Liben-Nowell and Jon M. Kleinberg. The link-prediction problem for social networks. *JASIST*, 58(7):1019–1031, 2007. DOI: 10.1002/asi.20591. 5, 11, 12, 13

[97] Ryan Lichtenwalter and Nitesh V. Chawla. Vertex collocation profiles: subgraph counting for link analysis and prediction. In *WWW*, pages 1019–1028, 2012. DOI: 10.1145/2187836.2187973. 13

[98] Ryan Lichtenwalter, Jake T. Lussier, and Nitesh V. Chawla. New perspectives and methods in link prediction. In *KDD'10*, pages 243–252, 2010. DOI: 10.1145/1835804.1835837. 13, 14

[99] Lu Liu, Jie Tang, Jiawei Han, and Shiqiang Yang. Learning influence from heterogeneous social networks. *Data Mining and Knowledge Discovery*, 25(3):511–544, 2012. DOI: 10.1007/s10618-012-0252-3. 7, 70

[100] Tiancheng Lou and Jie Tang. Mining structural hole spanners through information diffusion in social networks. In *WWW'13*, pages 837–848, 2013. 2, 3

[101] Tiancheng Lou, Jie Tang, John Hopcroft, Zhanpeng Fang, and Xiaowen Ding. Learning to predict reciprocity and triadic closure in social networks. *TKDD*, 2013, (accepted). DOI: 10.1145/2499907.2499908. 5, 11

[102] Sofus A. Macskassy and Foster Provost. A simple relational classifier. In *Workshop on Multi-Relational Data Mining in conjunction with KDD'03*, 2003. 100

[103] Shobhit Mathur, Marshall Scott Poole, Feniosky Peña-Mora, Mark Hasegawa-Johnson, and Noshir S. Contractor.   Detecting interaction links in a collaborating group using manually annotated data. *Social Networks*, 34:515–526, 2012. DOI: 10.1016/j.socnet.2012.04.002. 13, 15

[104] Julian McAuley and Jure Leskovec. Learning to discover social circles in ego networks. In *NIPS 2012 Workshop on Social Network and Social Media Analysis*, 2012. 171

[105] M. McPherson, L. Smith-Lovin, and J.M. Cook. Birds of a feather: Homophily in social networks. *Annual Review of Sociology*, pages 415–444, 2001. DOI: 10.1146/annurev.soc.27.1.415. 11

[106] Aditya Krishna Menon and Charles Elkan. A log-linear model with latent features for dyadic prediction. In *ICDM*, pages 364–373, 2010. DOI: 10.1109/ICDM.2010.148. 5

[107] Stanley Milgram. The small world problem. *Psychology Today*, 2:60–67, 1967. 1, 16

[108] Thomas Minka.   Estimating a dirichlet distribution.   In *Technique Report*, *http://research.microsoft.com/ minka/papers/dirichlet/*, 2003. 150, 153

[109] Thomas Minka and John Lafferty. Expectation-propagation for the generative aspect model. In *UAI'02*, pages 352–359, 2002. 150

[110] Bharath Kumar Mohan. The best nurturers in computer science research. In *SDM*, pages 566–570, 2005. 164

[111] Kevin P. Murphy, Yair Weiss, and Michael I. Jordan. Loopy belief propagation for approximate inference: An empirical study. In *UAI'99*, pages 467–475, 1999. 33, 61

[112] Seth A. Myers, Chenguang Zhu, and Jure Leskovec. Information diffusion and external influence in networks. In *KDD'12*, pages 33–41, 2012. DOI: 10.1145/2339530.2339540. 7

[113] M. E. J. Newman. Clustering and preferential attachment in growing networks. *Physical Review E*, 64(2):025102, 2001. DOI: 10.1103/PhysRevE.64.025102. 2

[114] M. E. J. Newman. Fast algorithm for detecting community structure in networks. *Physical Review E*, 69(066133), 2004. DOI: 10.1103/PhysRevE.69.066133. 2

[115] Satoshi Oyama, Kohei Hayashi, and Hisashi Kashima. Cross-temporal link prediction. In *ICDM*, pages 1188–1193, 2011. DOI: 10.1109/ICDM.2011.45. 13, 14, 19

[116] Lawrence Page, Sergey Brin, Rajeev Motwani, and Terry Winograd. The pagerank citation ranking: Bringing order to the web. Technical Report SIDL-WP-1999-0120, Stanford University, 1999. 56, 108

[117] Fragkiskoc Papadopoulos, Maksim Kitsak, M. Ángeles Serrano, Marián Bogu ná, and Dmitri Krioukov. Popularity versus Similarity in Growing Networks. *Nature*, 489:537–540, Sep 2012. DOI: 10.1038/nature11459. 13, 14

[118] Alexandrin Popescul and Lyle H. Ungar. Statistical relational learning for link prediction. In *IJCAI03 Workshop on Learning Statistical Models from Relational Data*, vol. 149, page 172, 2003. 6

[119] Tao Qin, Tie Y. Liu, Xu D. Zhang, De S. Wang, and Hang Li. Global ranking using continuous conditional random fields. In *NIPS*. MIT Press, 2008. 120

[120] Emile Richard, Nicolas Baskiotis, Theodoros Evgeniou, and Nicolas Vayatis. Link discovery using graph feature tracking. In *NIPS*, pages 1966–1974, 2010. 12, 13

[121] Matthew Richardson and Pedro Domingos. Mining knowledge-sharing sites for viral marketing. In *KDD'02*, pages 61–70, 2002. DOI: 10.1145/775047.775057. 7, 70, 85, 111

[122] Michal Rosen-Zvi, Thomas Griffiths, Mark Steyvers, and Padhraic Smyth. The author-topic model for authors and documents. In *UAI'04*, pages 487–494, 2004. 148, 155

[123] Maayan Roth, Assaf Ben-David, David Deutscher, Guy Flysher, Ilan Horn, Ari Leichtberg, Naty Leiser, Yossi Matias, and Ron Merom. Suggesting friends using the implicit social graph. In *KDD'10*, pages 233–242, 2010. DOI: 10.1145/1835804.1835836. 11

[124] Kazumi Saito, Ryohei Nakano, and Masahiro Kimura. Prediction of information diffusion probabilities for independent cascade model. In *KES '08*, pages 67–75, 2008. DOI: 10.1007/978-3-540-85567-5_9. 7

[125] Purnamrita Sarkar and Andrew W. Moore. Dynamic social network analysis using latent space models. *SIGKDD Exploration Newsletter*, 7(2):31–40, 2005. DOI: 10.1145/1117454.1117459. 8, 111

[126] Salvatore Scellato, Anastasios Noulas, and Cecilia Mascolo. Exploiting place features in link prediction on location-based social networks. In *KDD*, pages 1046–1054, 2011. DOI: 10.1145/2020408.2020575. 13

[127] Rossano Schifanella, Alain Barrat, Ciro Cattuto, Benjamin Markines, and Filippo Menczer. Folks in folksonomies: social link prediction from shared metadata. In *WSDM*, pages 271–280, 2010. DOI: 10.1145/1718487.1718521. 12, 13

[128] Jerry Scripps, Pang-Ning Tan, and Abdol-Hossein Esfahanian. Measuring the effects of preprocessing decisions and network forces in dynamic network analysis. In *KDD'2009*, pages 747–756, 2009. DOI: 10.1145/1557019.1557102. 7, 8, 111

[129] Burr Settles and Mark Craven. An analysis of active learning strategies for sequence labeling tasks. In *EMNLP*, pages 1070–1079, 2008. 41, 47

[130] Xiaolin Shi, Jun Zhu, Rui Cai, and Lei Zhang. User grouping behavior in online forums. In *KDD'09*, pages 777–786, 2009. DOI: 10.1145/1557019.1557105. 8, 111

[131] Xin Shuai, Ying Ding, Jerome Busemeyer, Shanshan Chen, Yuyin Sun, and Jie Tang. Modeling indirect influence on twitter. *IJSWIS*, 8(4):20–36, 2012. DOI: 10.4018/jswis.2012100102. 7

[132] George Simmel. *Sociological Theory*. 7th ed., McGraw Hill, New York, 2008. 1

[133] Parag Singla and Matthew Richardson. Yes, there is a correlation: - from social networks to personal behavior on the web. In *WWW'08*, pages 655–664, 2008. DOI: 10.1145/1367497.1367586. 6

[134] P. Smolensky. Information processing in dynamical systems: foundations of harmony theory. *Parallel Distributed Processing: Explorations in the Microstructure of Cognition, vol. 1: Foundations*, pages 194–281, 1986. 8

[135] Richard Sproat, Alan W. Black, Stanley Chen, Shankar Kumar, Mari Ostendorf, and Christopher Richards. Normalization of non-standard words. *Computer Speech Language*, pages 287–333, 2001. DOI: 10.1006/csla.2001.0169. 136

[136] Mark Steyvers, Padhraic Smyth, and Thomas Griffiths. Probabilistic author-topic models for information discovery. In *KDD'04*, pages 306–315, 2004. DOI: 10.1145/1014052.1014087. 148, 150, 155

[137] Chenhao Tan, Lillian Lee, Jie Tang, Long Jiang, Ming Zhou, and Ping Li. User-level sentiment analysis incorporating social networks. In *KDD'11*, pages 1397–1405, 2011. DOI: 10.1145/2020408.2020614. 8

[138] Chenhao Tan, Jie Tang, Jimeng Sun, Quan Lin, and Fengjiao Wang. Social action tracking via noise tolerant time-varying factor graphs. In *KDD'10*, pages 1049–1058, 2010. DOI: 10.1145/1835804.1835936. vi, 2, 8, 114

[139] Yee Fan Tan, Min-Yen Kan, and Dongwon Lee. Search engine driven author disambiguation. In *Proceedings of the 6th ACM/IEEE-CS Joint Conference on Digital Libraries (JCDL'06)*, pages 314–315, 2006. DOI: 10.1145/1141753.1141826. 145

[140] Jie Tang, A.C.M. Fong, Bo Wang, and Jing Zhang. A unified probabilistic framework for name disambiguation in digital library. *IEEE TKDE*, 24(6):975–987, 2012. DOI: 10.1109/TKDE.2011.13. 141, 145

[141] Jie Tang, Ruoming Jin, and Jing Zhang. A topic modeling approach and its integration into the random walk framework for academic search. In *ICDM'08*, pages 1055–1060, 2008. DOI: 10.1109/ICDM.2008.71. 81, 84, 85

[142] Jie Tang, Tiancheng Lou, and Jon Kleinberg. Inferring social ties across heterogeneous networks. In *WSDM'12*, pages 743–752, 2012. DOI: 10.1145/2124295.2124382. 5, 41, 172

[143] Jie Tang, Jimeng Sun, Chi Wang, and Zi Yang. Social influence analysis in large-scale networks. In *KDD'09*, pages 807–816, 2009. DOI: 10.1145/1557019.1557108. 2, 7, 70, 89, 102, 108, 109, 111

[144] Jie Tang, Sen Wu, and Jimeng Sun. Confluence: Conformity influence in large social networks. In *KDD'13*, pages 347–355, 2013. DOI: 10.1145/2487575.2487691. 111

[145] Jie Tang, Limin Yao, Duo Zhang, and Jing Zhang. A combination approach to web user profiling. *ACM TKDD*, 5(1):1–44, 2010. DOI: 10.1145/1870096.1870098. 134

[146] Jie Tang, Duo Zhang, and Limin Yao. Social network extraction of academic researchers. In *ICDM'07*, pages 292–301, 2007. DOI: 10.1109/ICDM.2007.30. 134

[147] Jie Tang, Jing Zhang, Ruoming Jin, Zi Yang, Keke Cai, Li Zhang, and Zhong Su. Topic level expertise search over heterogeneous networks. *Machine Learning Journal*, 82(2):211–237, 2011. DOI: 10.1007/s10994-010-5212-9. 153

[148] Jie Tang, Jing Zhang, Limin Yao, Juanzi Li, Li Zhang, and Zhong Su. Arnetminer: Extraction and mining of academic social networks. In *KDD'08*, pages 990–998, 2008. DOI: 10.1145/1401890.1402008. 2, 29, 34, 52, 82, 85, 123, 131

[149] Lei Tang and Huan Liu. Relational learning via latent social dimensions. In *KDD'09*, pages 817–826, 2009. DOI: 10.1145/1557019.1557109. 5, 111

[150] Wenbin Tang, Honglei Zhuang, and Jie Tang. Learning to infer social ties in large networks. In *ECML/PKDD'11*, pages 381–397, 2011. DOI: 10.1007/978-3-642-23808-6_25. 61

[151] Benjamin Taskar, Ming Fai Wong, Pieter Abbeel, and Daphne Koller. Link prediction in relational data. In *NIPS*, 2003. 6

[152] Yee Whye Teh, Michael I. Jordan, Matthew J. Beal, and David M. Blei. Hierarhical dirichlet processes. In *Technical Report 653,* Department of Statistics, UC Berkeley, 2004. 155

[153] C.J. van Rijsbergen. *Information Retrieval.* Butterworths, London, 1979. 133, 138

[154] Chao Wang, Venu Satuluri, and Srinivasan Parthasarathy. Local probabilistic models for link prediction. In *ICDM*, pages 322–331, 2007. DOI: 10.1109/ICDM.2007.108. 13, 14

[155] Chi Wang, Jiawei Han, Yuntao Jia, Jie Tang, Duo Zhang, Yintao Yu, and Jingyi Guo. Mining advisor-advisee relationships from research publication networks. In *KDD'10*, pages 203–212, 2010. DOI: 10.1145/1835804.1835833. 5, 13, 14, 25, 28, 29, 34, 37, 40, 52, 61

[156] Dashun Wang, Dino Pedreschi, Chaoming Song, Fosca Giannotti, and Albert-László Barabási. Human mobility, social ties, and link prediction. In *KDD*, pages 1100–1108, 2011. DOI: 10.1145/2020408.2020581. 13

[157] Stanley Wasserman and Katherine Faust. *Social Network Analysis: Methods and Applications.* Cambridge University Press, Cambridge, 1994. DOI: 10.1017/CBO9780511815478. 1

[158] Duncan J. Watts and Steven H. Strogatz. Collective dynamics of small-world networks. *Nature*, 440–442, 1998. DOI: 10.1038/30918. 2

[159] Max Weber. *The Nature of Social Action in Runciman, W.G. 'Weber: Selections in Translation'.* Cambridge University Press, 1991. 2, 111

[160] Xing Wei and W. Bruce Croft. Lda-based document models for ad-hoc retrieval. In *Proceedings of the 29th ACM SIGIR International Conference on Information Retrieval (SIGIR'06)*, pages 178–185, 2006. DOI: 10.1145/1148170.1148204. 154

[161] Gary M. Weiss. Mining with rarity: A unifying framework. *SIGKDD Explorations Newsletter*, pages 7–19, 2004. DOI: 10.1145/1007730.1007734. 20

[162] Max Welling and Geoffrey E. Hinton. A new learning algorithm for mean field boltzmann machines. In *Proceedings of International Conference on Artificial Neural Network (ICANN'01)*, pages 351–357, 2001. DOI: 10.1007/3-540-46084-5_57. 8

[163] Wim Wiegerinck. Variational approximations between mean field theory and the junction tree algorithm. In *UAI'00*, pages 626–633, 2000. 117

[164] Sen Wu, Jimeng Sun, and Jie Tang. Patent partner recommendation in enterprise social networks. In *WSDM'13*, pages 43–52, 2013. DOI: 10.1145/2433396.2433404. 13, 15

[165] Shaomei Wu, J. M. Hofman, W. A. Mason, and D. J. Watts. Who says what to whom on twitter. In *WWW'11*, pages 705–714, 2011. DOI: 10.1145/1963405.1963504. 54

[166] Endong Xun, Changning Huang, and Ming Zhou. A unified statistical model for the identification of english basenp. In *Proceedings of The 38th Annual Meeting of the Association for Computational Linguistics (ACL'00)*, pages 3–6, 2000. 136

[167] Yang Yang, Jie Tang, Cane Wing-ki Leung, Yizhou Sun, Qicong Chen, Juanzi Li, and Qiang Yang. Rain: Social role-aware information diffusion. In *AAAI'15*, 2015. 7

[168] Zi Yang, Jingyi Guo, Keke Cai, Jie Tang, Juanzi Li, Li Zhang, and Zhong Su. Understanding retweeting behaviors in social networks. In *CIKM'10*, pages 1633–1636, 2010. DOI: 10.1145/1871437.1871691. 41

[169] Zi Yang, Jie Tang, Bo Wang, Jingyi Guo, Juanzi Li, and Songcan Chen. Expert2bole: From expert finding to bole search. In *Proceeding of the 15th ACM SIGKDD International Conference on Knowledge Discovery and Data Mining (KDD'09)*, 2009. 164

[170] Jonathan S. Yedidia, William T. Freeman, and Yair Weiss. Generalized belief propagation. In *NIPS'01*, pages 689–695, 2001. 117

[171] Xiaoxin Yin, Jiawei Han, and Philips Yu. Object distinction: Distinguishing objects with identical names. In *Proceedings of IEEE 23rd International Conference on Data Engineering (ICDE'2007)*, pages 1242–1246, 2007. DOI: 10.1109/ICDE.2007.368983. 145, 146

[172] Zhijun Yin, Manish Gupta, Tim Weninger, and Jiawei Han. Linkrec: a unified framework for link recommendation with user attributes and graph structure. In *WWW*, pages 1211–1212, 2010. DOI: 10.1145/1772690.1772879. 13, 14

[173] Chengxiang Zhai and John Lafferty. A study of smoothing methods for language models applied to ad hoc information retrieval. In *SIGIR'01*, pages 334–342, 2001. DOI: 10.1145/383952.384019. 154, 155

[174] Jing Zhang, Biao Liu, Jie Tang, Ting Chen, and Juanzi Li. Social influence locality for modeling retweeting behaviors. In *IJCAI'13*, pages 2761–2767, 2013. 7, 171

[175] Jing Zhang, Jie Tang, and Juanzi Li. Expert finding in a social network. In *DASFAA'07*, pages 1066–1069, 2007. DOI: 10.1007/978-3-540-71703-4_106. 153

[176] Jun Zhang, Chaokun Wang, Philip S. Yu, and Jianmin Wang. Learning latent friendship propagation networks with interest awareness for link prediction. In *SIGIR'13*, pages 63–72, 2013. DOI: 10.1145/2484028.2484029. 13, 15

[177] Jun Zhu. Max-margin nonparametric latent feature models for link prediction. In *ICML*, pages 719–726, 2012. 13, 14

[178] Yaojia Zhu, Xiaoran Yan, Lise Getoor, and Cristopher Moore. Scalable text and link analysis with mixed-topic link models. In *KDD*, pages 473–481, 2013. DOI: 10.1145/2487575.2487693. 13, 15

Authors' Biographies

JIE TANG

Jie Tang is an associate professor with the Department of Computer Science and Technology, at Tsinghua University. His interests include social network analysis, data mining, and machine learning. He has published more than 100 journal/conference papers and holds 10 patents. He served as PC Co-Chair of WSDM'15, ASONAM'15, ADMA'11, SocInfo'12, KDD-CUP Co-Chair of KDD'15, Poster Co-Chair of KDD'14, Workshop Co-Chair of KDD'13, Local Chair of KDD'12, Publication Co-Chair of KDD'11, and as the PC member of more than 50 international conferences. He is the principal investigator of National High-tech R&D Program (863), NSFC project, Chinese Young Faculty Research Funding, National 985 funding, and international collaborative projects with Minnesota University, IBM, Google, Nokia, Sogou, etc. He leads the project Arnetminer.org for academic social network analysis and mining, which has attracted millions of independent IP accesses from 220 countries/regions in the world. He was honored with the Newton Advanced Scholarship Award, CCF Young Scientist Award, NSFC Excellent Young Scholar, and IBM Innovation Faculty Award.

JUANZI LI

Juanzi Li obtained her Ph.D. degree from Tsinghua University. She is now a full professor at Tsinghua University. Her main research interest is to study the semantic technologies by combining the Natural Language Processing, Semantic Web and Data Mining. She is the vice director of Chinese Information Processing Society of Chinese Computer Federation in China. She is the principal investigators of many key projects supported by the Natural Science Foundation of China (NSFC), the national basic science research program and international cooperation projects. She has published over 120 papers in many international journals and conferences such as TKDE, SIGIR, SIGMOD, SIGKDD, IJCAI, etc.

Printed in the United States
by Baker & Taylor Publisher Services